Bitingduck press

Truly Tricky Graduate Physics Problems

with Solutions

Dr. Jay Nadeau
Dr. Ben Sauerwine
Leila Cohen

Preface

This book is a labor of love. We wrote it because we loved solving physics problems as grad students, but sometimes got frustrated when it was hard to obtain detailed solutions to find out if we were on the right track. Here we've tried to collect the most representative problems from the 5 major branches of physics–classical mechanics, special relativity, quantum mechanics, statistical/thermal physics, and electricity and magnetism–and present the solutions in great detail, pointing out when there are special tricks or leaps of logic.

Many of the "classic" problems may be found in the literature, where they were new and original when they were solved for the first time. In these cases, we provide references to these original papers, and encourage you to read them. Sometimes a problem can be solved in more than one way, and the original solutions can be surprisingly elegant or complicated, as the case may be.

The majority of these problems can be solved quickly once the tricks are seen.

They are the type of problem that is likely to be found on a midterm or qualifying exam. A few are longer (e.g. Aharonov-Bohm effect) and are of the type seen in take-home exams or homeworks. Our primary audience is graduate students studying for qualifying exams, although junior/senior undergraduates will also find useful problems here.

We have made every effort to eliminate typos and other errors, but no doubt some remain. We encourage you to contact us with errata. Please note that if this book's other formats (such as e-pub) are used, browsers may render differently. One known problem is that in the conversion from epub to pdf using Calibre, some of the square brackets [] get lost. Some browsers cannot render certain characters, or may show them differently. If something seems wrong in these versions, we encourage you to consult the html or use another reader or browser to check the equation.

Happy solving!

Published by Bitingduck Press

Copyright 2013 Nadeau,
Cohen,Sauerwine

ISBN 978-1-938463-17-4 (print)

978-1-938463-01-3 (electronic)

For information contact:
Bitingduck Press, LLC

Altadena · Montreal

www.bitingduckpress.com

Cover image: Mechanical Duck,
built by Jacques de Vaucanson
(1738, France)

Image from Wikimedia Commons
(public domain)

List of Problems

Chapter 1: Short Answer

Chapter 2: Classical Mechanics

Problem 2.01: Coriolis deflection

Problem 2.02: Water swirling down the drain vs. latitude

Problem 2.03: Sphere rolling off another sphere

Problem 2.04: Frequency of small oscillations

Problem 2.05: Circular orbits

Problem 2.06: Off-center circular orbit

Problem 2.07: Escape velocity

Problem 2.08: Spherical pendulum

Problem 2.09: Atwood machine

Problem 2.10: Moments of inertia

Problem 2.11: A yo-yo

Problem 2.12: Slingshot maneuver

Problem 2.13: Ball in a bucket

Problem 2.14: Front-wheel versus rear-wheel drive

Problem 2.15: Planet in a dust cloud

Problem 2.16: Pendulum on a sliding support

Problem 2.17: Enmeshed gears

Problem 2.18: Bead on a hoop

Problem 2.19: Ball falling through the earth

Problem 2.20: Star in the gravitational potential of a galaxy

Problem 2.21: Mass between two springs

Problem 2.22: Double pendulum

Problem 2.23: Cylindrical force

Problem 2.24: Spherical ball with varying density on a slope

Problem 2.25: Pendulum on a sled

Chapter 3: Special Relativity

Problem 3.01: Compton scattering

Problem 3.02: Compton scattering at right angles

Problem 3.03: Compton scattering maximum angle

Problem 3.04: Phase velocity of an electron

Problem 3.05: Antiproton detection

Problem 3.06: Decay of unknown particle

Problem 3.07: Galactic red-shift

Problem 3.08: Muons on Mt. Washington

Problem 3.09: Upper bound on photon mass

Problem 3.10: Photon mass, another approach

Problem 3.11: Photon colliding with a nucleus

Problem 3.12: Relativistic time dilation

Problem 3.13: Mossbauer spectroscopy

Problem 3.14: Thermal photon collisions in space

Problem 3.15: Relativistic effects on hydrogen atom

Problem 3.16: Relativistically moving rod

Problem 3.17: Precession of sun's perihelion

Chapter 4: Quantum Mechanics

Problem 4.01: Finite nuclear size

Problem 4.02: Geiger-Nuttal relation

Problem 4.03: Atom in a magnetic field

Problem 4.04: System of fermions

Problem 4.05: Kronig-Penney model

Problem 4.06: Harmonic oscillator in an E field

Problem 4.07: Aharonov-Bohm effect

Problem 4.08: Ensemble of neutrons

Problem 4.09: Spin matrices of a spin-1 particle

Problem 4.10: State vector of spin-½ particles

Problem 4.11: Density matrix of spin-½ particles

Problem 4.12: Ensemble of spin-1 particles

Problem 4.13: Raising and lowering operators

Problem 4.14: System with j = 1

Problem 4.15: Deuteron

Problem 4.16: Circular potential

Problem 4.17: 3D harmonic oscillator

Problem 4.18: Expanding square well potential

Problem 4.19: Square well potential and Heisenberg

Problem 4.20: Asymmetric box

Problem 4.21: Two particles, one box

Problem 4.22: Nuclear quadrupoles and NMR

Problem 4.23: Born-Oppenheimer approximation

Problem 4.24: The hydrogen molecule

Problem 4.25: Rotational transitions of a diatomic molecule

Problem 4.26: Vibrational transitions of a diatomic molecule

Problem 4.27: Lennard-Jones

Problem 4.28: Rotational absorption lines

Problem 4.29: The KCl molecule

Problem 4.30: Ramsauer-Townsend effect

Problem 4.31: Phase shift from a scattering potential

Problem 4.32: A special operator

Problem 4.33: Energy levels of carbon

Problem 4.34: An interesting wave function

Problem 4.35: The helium atom

Problem 4.36: Highly excited hydrogen

Problem 4.37: Pulling apart metal

Problem 4.38: Photoelectric emission

Problem 4.39: Spin-½ system

Problem 4.40: Beta decay of a nucleus

Problem 4.41: Yukawa potential

Problem 4.42: Transmission through a barrier

Problem 4.43: Particle on a surface

Problem 4.44: Electron on a conducting surface

Problem 4.45: Bosons on a ring

Problem 4.46: Backbending phenomenon

Problem 4.47: Fine and ultrafine

Chapter 5: Statistical and Thermal Physics

Problem 5.01: Entropy definition

Problem 5.02: Stirling approximation

Problem 5.03: Grand Canonical ensemble

Problem 5.04: 2D Fermi gas

Problem 5.05: Separation of gases by a piston

Problem 5.06: Ultrarelativistic Fermi gas

Problem 5.07: Ultrarelativistic Bose gas

Problem 5.08: Bose-Einstein condensation

Problem 5.09: Speed of sound

Problem 5.10: Entropy per particle in a quantum gas

Problem 5.11: Dipole gas in an electric field

Problem 5.12: Van der Waals equation

Problem 5.13: Specific heat of van der Waals gas

Problem 5.14: Van der Waals gas critical point

Problem 5.15: Quantum gas in classical limit

Problem 5.16: Helium λ temperature

Problem 5.17: Why no Fermi condensation?

Problem 5.18: Ergodic systems

Problem 5.19: Quantum Hall effect

Problem 5.20: Chandrasekhar limit

Problem 5.21: Plasma frequency

Problem 5.22: Free-electron theory of metals

Problem 5.23: Doped semiconductor

Problem 5.24: Diatomic molecule

Problem 5.25: Moon's atmosphere

Problem 5.26: Magnetization of Fermi gas

Problem 5.27: Entropy of a polymer

Problem 5.28: Negative temperature

Problem 5.29: Clausius-Clapeyron equation

Problem 5.30: Fahrenheit xxx

Problem 5.31: Pumping on helium

Problem 5.32: Effusion

Problem 5.33: Carnot engine

Problem 3.34: Magnetic system

Problem 5.35: Photonic cavity

Problem 5.36: Frenkel defects

Problem 5.37: Filling a balloon

Problem 5.38: Could the Sun be heated by gravity?

Problem 5.39: Stefan-Boltzmann law

Problem 5.40: Atmosphere of Neptune

Problem 5.41: Heating the house

Problem 5.42: Vacuum system

Problem 5.43: Blackbody in a cavity

Problem 5.44: Five-state system

Problem 5.45: Atoms in a sticky box

Problem 5.46: A simple crystal

Problem 5.47: Method of Jacobians

Problem 5.48: Jacobians II

Problem 5.49: A two-phase material

Problem 5.50: Four spin-1 particles

Problem 5.51: Hypersphere

Problem 5.52: Ultrarelativistic gas

Problem 5.53: Anharmonic oscillator

Chapter 6: Electricity and Magnetism

Problem 6.01: Conducting sphere in an electric field

Problem 6.02: Polarized spherical shell

Problem 6.03: Charged spherical shell

Problem 6.04: One charged sphere inside another

Problem 6.05: Conducting sphere with a hole

Problem 6.06: Iron rod wound with wire

Problem 6.07: Stokes parameters

Problem 6.08: Plane wave reflected and transmitted by dielectric film

Problem 6.09: H- ions in a cyclotron

Problem 6.10: Parallel plate capacitor with electric field

Problem 6.11: Circularly polarized plane wave scattered by a free electron

Problem 6.12: The Larmor formula

Problem 6.13: Power radiated by a charge in orbit

Problem 6.14: Rotating dipole

Problem 6.15: Point charges on a rotating circle

Problem 6.16: Instability of classical hydrogen atom

Problem 6.17: Rotating bar in magnetic field

Problem 6.18: Circular loop in magnetic field

Problem 6.19: Frame falling through magnetic field

Problem 6.20: Radiation pressure on the Earth from the Sun

Problem 6.21: Poynting-Robertson effect

Problem 6.22: Ladder circuit

Problem 6.23: Meissner effect

Problem 6.24: Electron falling through a charged loop

Problem 6.25: Point charge above a conducting plane

Problem 6.26: Skin depth

Problem 6.27: Dipole radiation from a sphere

Problem 6.28: Alternating-circuit capacitor

Problem 6.29: Sphere with radially varying charge density

Problem 6.30: Index of refraction

Problem 6.31: Spacelike 4-vector and vector potential

Problem 6.32: Equation of continuity

Problem 6.33: Mean value theorem

Problem 6.34: Vector potential of a wire

Problem 6.35: Wire loop with alternating current

Problem 6.36: Two oscillating dipoles

Problem 6.37: Dipole and quadrupole moments of a line charge

Problem 6.38: A special vector potential

Problem 6.39: Energy of two dipoles

Problem 6.40: Magnetic field of spinning sphere

Problem 6.41: Relativistic particle in electric and magnetic fields

Problem 6.42: Relativistic particle in a static potential

1

General knowledge

Problems

General knowledge

1. Discuss the difference between fermions and bosons.

2. What are baryons and mesons? How do they differ?

3. What is the classical electron radius?

4. What is the highest kinetic energy that has been attained at a particle accelerator? (Bonus: what facility?)

5. What is the mean lifetime of a free neutron? How does it decay?

6. What is the expected possible decay mechanism of the proton? The expected lifetime?

7. At one atmosphere of pressure, what are the temperatures of liquid nitrogen and liquid helium?

8. What is the average speed of a nitrogen molecule in this room?

9. What would be a typical mean lifetime for an excited atom that has an allowed transition to the ground state with emission of a visible photon?

10. What is the mass of the muon relative to that of the electron?

11. Approximately how much power (in W/m^2) arrives at the Earth's surface on a sunny day in Los Angeles? How about a sunny day in Barrow, Alaska (71 degrees N latitude)?

12. What is the typical angular resolution (in minutes) of the human eye?

13. How does the lattice heat capacity of graphite scale with temperature?

14. What is the orbital speed of Neptune relative to that of the Earth?

15. What are the typical wavelengths of the following?

    a. radio waves

    b. infrared light

    c. visible light

    d. ultraviolet light

    e. x-rays

    f. gamma rays

16. Define "radiation." What is "ionizing radiation?" What is "radioactivity?"

17. If the Universe consisted of the Earth and the Sun, what would an observer at the center of the Sun say about the Earth's orbital shape and location of the foci? What about an observer in the center of the Earth? An observer at the center of mass?

18. Give the dimensions of the following:

    a. frequency

    b. entropy

    c. moment of inertia

    d. gravitational constant

    e. Poynting vector

19. Give the following in terms of basic physical constants:

    a. Compton wavelength

    b. de Broglie wavelength

    c. ground state of a hydrogen atom

    d. first Bohr radius of a hydrogen atom

    e. rest energy of an electron

    f. energy of a photon

    g. fine structure constant

    h. Hall coefficient

20. When do the highest tides occur, and why?

21. What is synchrotron radiation? Cerenkov radiation?

22. What is a neutron star? What keeps it from collapsing into a black hole?

23. What are the eigenvalues of an $n \times n$ matrix where all elements are equal to 1?

24. How does the wind resistance scale with velocity? How does this affect a rider in the Tour de France?

25. What happens to someone who falls into lava?

26. Define order and chaos.

27. What quarks make up a proton? A neutron?

28. How were neutrinos discovered?

29. What is the Stark effect? The Zeeman effect?

30. What is a negative temperature?

31. What are the major components of cosmic rays?

32. What was the first antiparticle observed?

33. What ice melts in water, does the water line go up or down?

34. Why can't He-4 be solidified?

35. What is the velocity of a nucleon at the Fermi surface? How long does it take the particle to cross a nucleus of $A = 100$?

36. Estimate the time and temperature at which nucleons froze out after the Big Bang.

37. Estimate the n-to-p ratio in the early Universe at a temperature of $T = 10^{10}$ K.

38. The vector bosons of the weak interaction are the W bosons with mass $m_w = 81$ GeV. What is the predicted range of this force?

39. Give typical values for the following:

    a. the Fermi energy of liquid ^{3}He

    b. the Fermi energy of a metal

    c. the Debye temperature of a crystalline solid

    d. the separation between the ground and first excited rotational states of a diatomic gas

    e. the energy of the first excited electronic state of a diatomic gas

40. Give values for the following:

    a. the diameter of a gold atom

    b. the ratio of the pion mass to the neutron mass

    c. the temperature of cosmic background radiation

    d. the diameter of a carbon nucleus

    e. the speed of sound in air

    f. the average binding energy in a nucleus

2

Classical Mechanics

Problems

Classical Mechanics

Problem 2.01

Solution 2.01

At a spot on earth at latitude L, a ball is thrown upward with initial velocity v_0. By the time it falls, how much is it deflected by the rotation of the earth? Can you think of a real-life event where this effect is or was noticeable?

Problem 2.02

Solution 2.02

It is commonly believed that water swirls down a drain in opposite directions in the Northern and Southern hemispheres. Does this make sense? Justify your answer.

Problem 2.03

Solution 2.03

A small sphere (radius R_2) is balanced on a large sphere (radius R_1) and begins rolling off. At what angle from the vertical does it leave the larger sphere? What if the small sphere is a point mass?

Problem 2.04

Solution 2.04

A particle of mass m is trapped between two other particles of mass M that both repel it with a potential:

$$U = \frac{mM}{r^n}$$

Find the frequency of small oscillations of the center mass along the line of the particles, if the total distance along the line is $2d$, $n = 1,2$, etc.

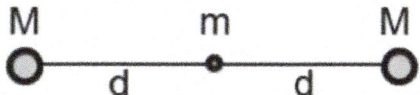

Problem 2.05

Solution 2.05

Given an attractive force $F = cr^n$, for which values of n is a stable circular orbit possible?

Problem 2.06

Solution 2.06

If a particle in a circular orbit has an orbit that passes through the point of attraction, show that the force must satisfy:

$$F = \frac{c}{r^5}$$

Problem 2.07

Solution 2.07

Derive an expression for the escape velocity from a planet. If the planet is Earth, give a numerical value using $m_e = 6 \times 10^{24}$ kg, $r_e = 6.4 \times 10^6$ m.

Problem 2.08

Solution 2.08

Write the Lagrangian and equations of motion for a spherical pendulum: that is, a mass m attached to a string of length l, free to move in three dimensions.

Problem 2.09

Solution 2.09

Write the Lagrangian and equations of motion for the Atwood machine (pictured; the pulley has radius R and moment of inertia I).

Friction-less pulley

m2

m1

Problem 2.10

Solution 2.10

Give the expression for the rotational kinetic energy of a 3-dimensional body. Under what circumstances is it a maximum?

Problem 2.11

Solution 2.11

A yo-yo can be considered as two disks of radius R connected by a massless cylinder of radius r around which the string is wrapped. If the upper end of the string is held fixed, find the downward acceleration of the yo-yo after it is released, and the tension in the string.

Problem 2.12

Solution 2.12

Spacecraft can be sent to the outer planets by performing a "slingshot" maneuver around Jupiter. A rocket of mass m and velocity v travels in a trajectory perpendicular to Jupiter's orbit, then interacts gravitationally and by conservation of energy, ends up with a new velocity at some angle to Jupiter's orbit. Find an expression for this new velocity, assuming that Jupiter's orbit remains unchanged.

Problem 2.13

Solution 2.13

A plastic ball of density < 1 g/mL is in a water-filled bucket, attached to the bottom of the bucket by a stretched spring. If the bucket is dropped, does the ball rise, sink, or stay in the same position relative to the water?

Problem 2.14

Solution 2.14

Think of an automobile as two axles a distance L apart, with a center of gravity located between them and a distance h off the ground.

a. Calculate the steepest hill the car can climb in the case of front-wheel drive versus rear-wheel drive. Assume the coefficient of friction is known.

b. Calculate the maximum acceleration possible under both conditions.

Problem 2.15

Solution 2.15

A satellite orbits a planet at a distance R. The planet's mass is M and its radius is R_0, and it is also surrounded by a uniform cloud of dust extending out far beyond the satellite's orbit to a distance R_1. Assuming the dust does not impact the satellite's motion, calculate the gravitational potential at R; the period of a circular orbit; and the period of small oscillations about the circular orbit.

Problem 2.16

Solution 2.16

A one-dimensional pendulum of mass m, length l is attached to a support which moves horizontally back and forth according to $x = A\cos(\omega t)$. Write the Lagrangian of this system and discuss the limiting condition of small pendulum displacements.

Problem 2.17

Solution 2.17

Two turning gears of different radii and moments of inertia are suddenly enmeshed. Discuss the change in angular velocity of the gears and the conservation of angular momentum.

Problem 2.18

Solution 2.18

A bead slides on a frictionless hoop that is swinging in a vertical plane about a fixed pivot point. Write the Lagrangian and find the frequencies of normal modes for small oscillations as a function of the inclination of the hoop to the vertical θ and the inclination of the bead ϕ.

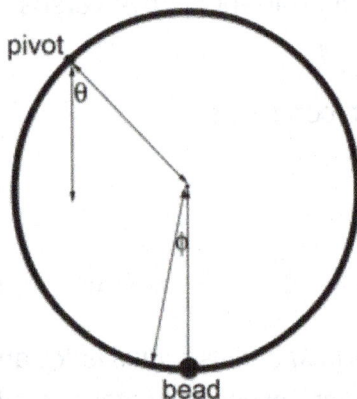

Problem 2.19

Solution 2.19

A ball of mass m is dropped through a hole drilled through the earth (mass M_E, radius R_E; assume constant density). Describe the motion and compare the period to that of a circular orbit just above the Earth's surface.

Problem 2.20

Solution 2.20

The motion of a star through a disk galaxy can be modeled as a point mass m released from rest at a distance d above a disk of radius R and thickness L, where $L \ll d \ll R$. The disk has a uniform density and a total mass $M \gg m$. Describe the motion.

Problem 2.21

Solution 2.21

A mass is placed on a rough surface between two springs of spring constant k. The frictional force opposing the motion is μv. Write and solve the differential equation for the motion.

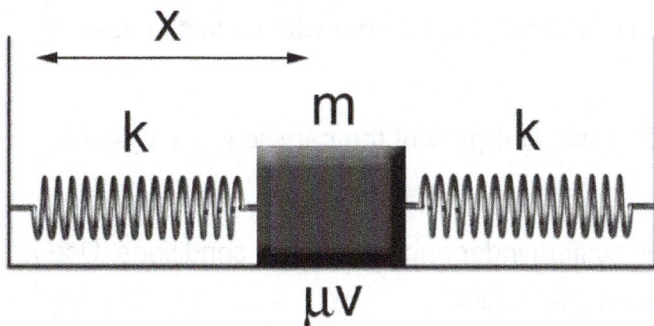

Problem 2.22

Solution 2.22

Consider small planar oscillations of a double pendulum consisting of two metal bars, the upper of mass $2m$ and length D and the lower one of mass m and length $2D$. Each bar has uniform mass density and negligible width and thickness. Let θ_1 denote the angle of the upper bar from the vertical and θ_2 denote the angle of the lower bar from the vertical. All frictional effects are negligible. The moment of inertia about an axis perpendicular to and passing through the center of a long thin rod of mass M and length L is $\frac{1}{12}ML^2$.

a. Determine the characteristic angular frequencies ω_1 and ω_2 of the normal modes of small oscillation. Ensure $\omega_1 < \omega_2$.

b. Determine the normal coordinates η_1 and η_2 corresponding to ω_1 and ω_2, respectively. Express these in terms of θ_1 and θ_2, ignoring overall constants.

Problem 2.23

Solution 2.23

A particle of mass m is acted upon by a constant gravitational field with acceleration g. The particle is also acted upon by a force $F = kr$ that is directed toward the origin, where r is the distance from the particle to the origin. There is a cylinder of radius R into which the particle cannot enter. The particle is constrained to slide on the cylinder with no friction (see the figure on the following page).

a. Write the kinetic energy K and the potential energy V of the particle in the cylindrical coordinate system: (ρ, ϕ, z).

b. Write the equations of motion for this system independent of initial conditions. Use F_{cyl} to represent the force of the cylinder on this particle.

c. Calculate the magnitude of F_{cyl} on the particle.

d. Identify all relevant conservation laws that are valid for the motion of the particle when it remains on the surface of the cylinder. Justify why the laws you have identified apply to this specific motion.

e. Given the initial conditions at $t = 0$: $z_0 = 0, \dot{z}_0 = 0, \phi_0 = 0, \dot{\phi}_0 \neq 0$, determine the position of the particle on the surface of the cylinder at any subsequent time.

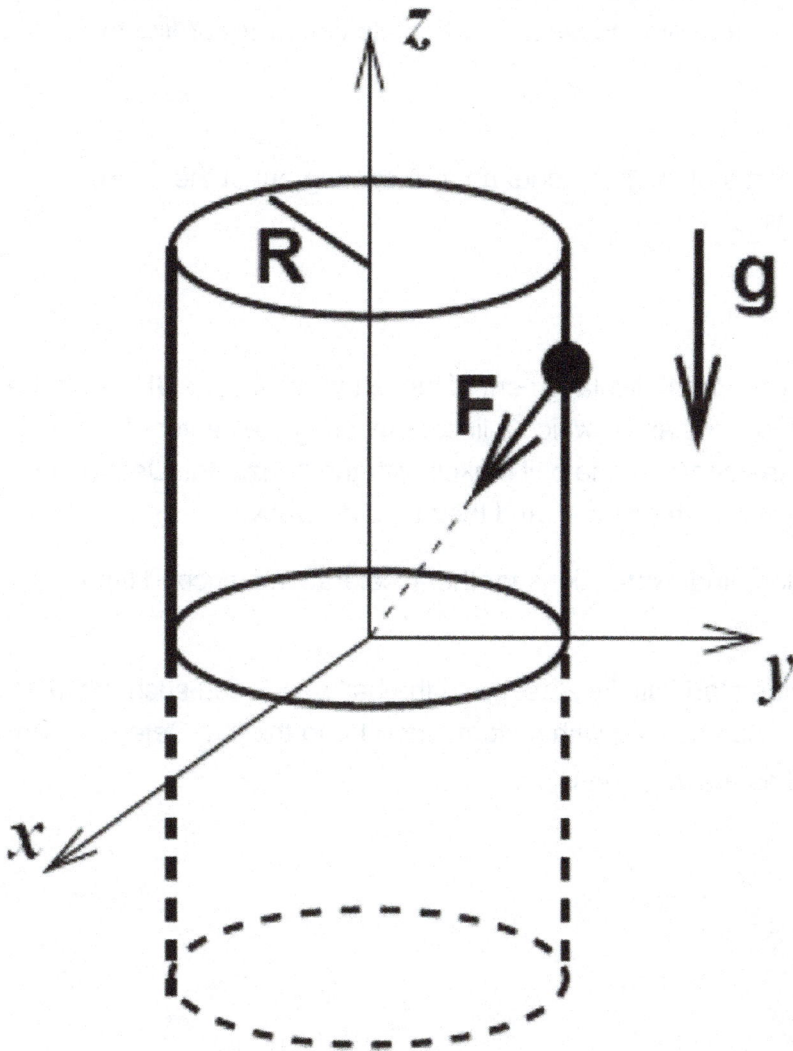

Problem 2.24

A spherical ball of radius a and mass m which is not uniformly distributed throughout its volume rolls without slipping down the circular surface of radius of curvature $4a$ of a wedge of mass $5m$ which is free to slide frictionlessly across a smooth horizontal surface, as shown. The density of the ball varies with the distance r from its center according to $\frac{\rho(r)}{\rho(0)} = 1 - \frac{4r}{5a}$.

a. Show that the moment of inertia of the ball about an axis coming out of the page through the center of the ball is

$$I = \frac{1}{3}ma^2$$

b. Let x denote the rightward horizontal displacement of the left vertical edge of the wedge from the origin O, and θ denote the angle which a line connecting the center of curvature of the surface to the center of the ball makes with the horizontal. Determine the Lagrangian of the system in terms of x, θ, and their time derivatives.

c. From the equations of motion, find expressions for the accelerations (second time derivatives of x and θ).

d. If both the ball and the wedge start initially at rest with the ball positioned such that $\theta = 0$, eventually the wedge ends up moving with uniform speed v_f to the left. Determine an expression for v_f, simplified as much as possible.

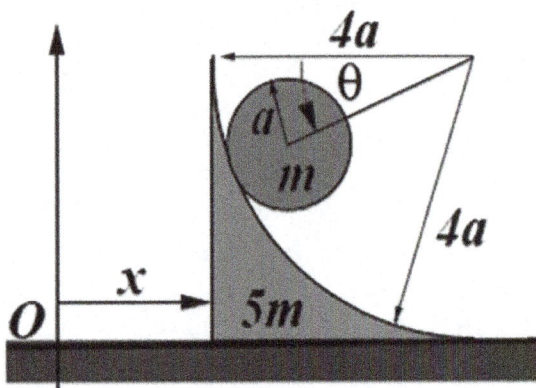

Problem 2.25

A pendulum of mass m and length b is mounted to a heavy sled of mass βm , which is free to slide with negligible friction down an inclined plane of slope α, as shown. The entire system acts under the influence of gravity, where g is the downward acceleration due to gravity.

a. Let ϕ denote the angle that the pendulum bob makes with the vertical and s denote the distance that the sled has slid down the slope (see diagram). Determine the Lagrangian describing the system in terms of this angle and distance and their time derivatives ($L(s,\phi,\dot{s},\dot{\phi})$).

b. Find expressions for the conjugate momenta p_s , p_ϕ.

c. Describe the steps you would use to find the Hamiltonian of the system. Do not actually compute H.

d. From the equations of motion for the system find expressions for the generalized accelerations (second time derivatives of s and ϕ).

e. The sled can slide down the slope with the pendulum at an unchanging angle ϕ_0 if ϕ_0 satisfies a certain condition. What is this condition?

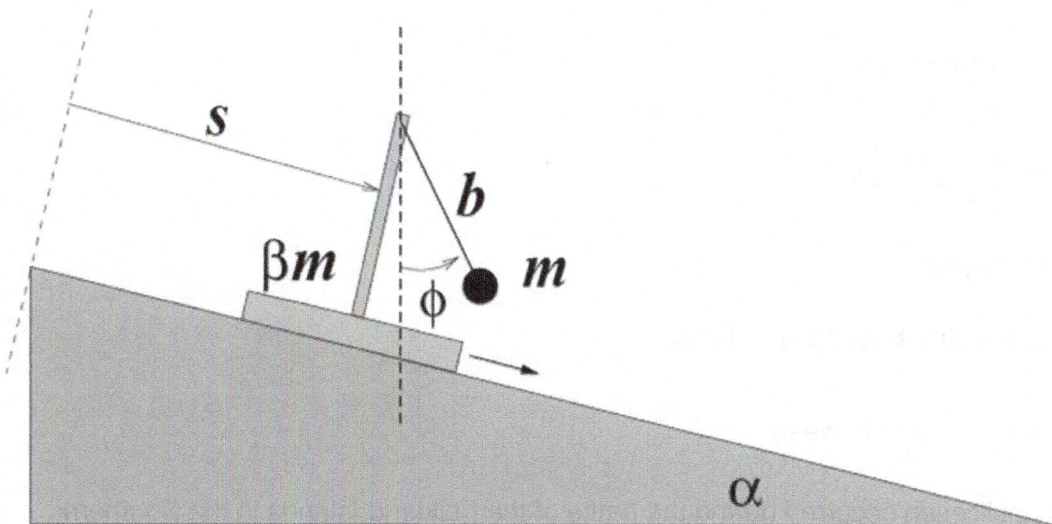

Solutions
Classical Mechanics

Solution 2.01

Problem 2.01

The ball first travels up, then down, with time of flight $t = \frac{v_0}{g}$ in each direction. The deflection will be the same in each; just multiply by 2. Define the $-y$ axis as west, and $+y$ as east; the z axis is up. Then:

$$F = -mg\hat{z} - 2m\vec{\omega} \times \vec{v} = -mg\hat{z} - 2mv\omega\cos\theta\,\hat{y}$$

$$\frac{d^2y}{dt^2} = -2v\omega\cos\theta$$

$$\frac{d^2y}{dt^2} = -g$$

$$v = v_0 - gt$$

$$\Rightarrow \frac{d^2y}{dt^2} = -2\omega\cos\theta(v_0 - gt)$$

$$\frac{dy}{dt} = -2\omega\cos\theta\left(v_0 t - \frac{1}{2}gt^2\right)$$

$$y = -2\omega\cos\theta\left(\frac{1}{2}v_0 t^2 - \frac{1}{3}gt^3\right)$$

Substitute $t = v_0/g$:

$$y = -2\omega\cos\theta\frac{v_0^3}{3g^2} \text{ (for both up and down)}$$

$$y_{\text{total}} = -\frac{4}{3}\omega\cos\theta\frac{v_0^3}{g^2} \text{ (to the west)}$$

A real-life application occurred during the battle of the Falkland Islands in the Southern hemisphere. The guns were calibrated for the Northern Hemisphere, where $\theta > 0$; when they were fired in the Southern Hemisphere, they landed many meters away from their target.

Solution 2.02 Problem 2.02

It doesn't make sense. The Coriolis force is very small and is readily overwhelmed by asymmetry in the drain itself, and vorticity added when the water is introduced or splashed around.

To put in perspective:

$$a = -2\omega v \cos\theta$$

$$\omega = \frac{360 \text{ deg}}{24 \text{ hr}}$$

$v \approx 50$ m/s for a strong weather system

$a \approx 10 \ \mu$m/s

So, even for a hurricane, it takes hours to days for the effect to become apparent. Water swirling down a drain goes too fast for the Coriolis effect to be seen.

In order to observe the Coriolis force in a pan of water, it must be a perfectly symmetric pan that has been allowed to sit for days in order to dissipate residual vorticity; the plug must be pulled in a perfectly symmetric fashion. Also note that many "Coriolis scams" involve people claiming that just crossing the equator suddenly reverses the direction of water spinning— since just stepping over the equator will be at latitude approximately 0 in both cases, the effect will be vanishingly small. Hurricanes do not form within 5 degrees of the equator because of insufficient Coriolis rotation.

This is also why a Foucault pendulum must be set in place with a thread; the thread is then burned— any spin or wobble introduced to the motion would overwhelm the Coriolis force.

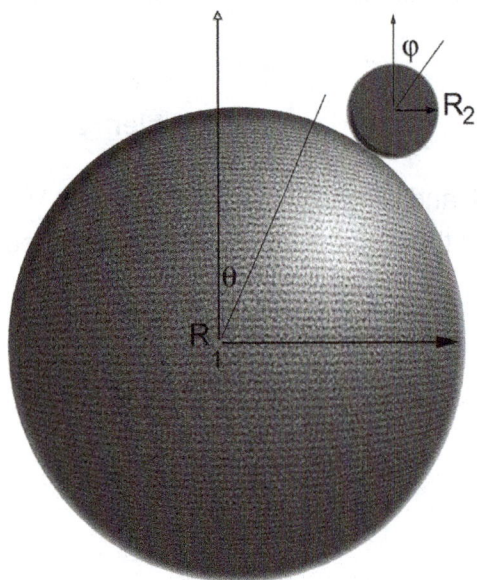

Write the Lagrangian with an equation of constraint:

$$L = \frac{1}{2}m\dot{r}^2 + \frac{1}{2}mr^2\dot{\theta}^2 + \frac{1}{2}I\dot{\varphi}^2 - mgr\cos\theta$$

$$(R_1 + R_2)\theta = R_2\varphi$$

$$\Rightarrow L = \frac{1}{2}m\dot{r}^2 + \frac{1}{2}mr^2\dot{\theta}^2 + I\left(\frac{R_1 + R_2}{R_2}\right)^2\dot{\theta}^2 - mgr\cos\theta$$

This gives:

$$\frac{\partial L}{\partial r} = mr^2\dot{\theta}^2 - mg\cos\theta$$

$$\frac{\partial L}{\partial \dot{r}} = m\dot{r}$$

$$\frac{\partial L}{\partial \theta} = mgr\sin\theta$$

$$\frac{\partial L}{\partial \dot{r}} = mr^2\dot{\theta} + I\left(\frac{R_1 + R_2}{R_2}\right)^2 \dot{\theta}$$

$$\Rightarrow \ddot{r} = r\dot{\theta}^2 - g\cos\theta$$

$$\ddot{\theta}\left[mr^2 + I\left(\frac{R_1 + R_2}{R_2}\right)^2\right] = mgr\sin\theta$$

The small sphere leaves the large one when:

$$\dot{r} = \ddot{r} = 0$$

$$r = R_1 + R_2$$

Plug these in to the equations of motion to give:

$$(R_1 + R_2)\dot{\theta}^2 = g\cos\theta \text{ (Eq. 1)}$$

$$\ddot{\theta} = \frac{d\theta}{dt}\frac{d}{d\theta}\frac{d\theta}{dt} = \dot{\theta}\frac{d\dot{\theta}}{d\theta}$$

$$\int_\theta^{\dot{\theta}} \dot{\theta} d\dot{\theta} = \int_0^\theta \frac{mg}{(R_1 + R_2)\left(m + \frac{I}{R_2^2}\right)}\sin\theta d\theta$$

$$\frac{1}{2}\dot{\theta}^2 = \frac{mg}{(R_1 + R_2)\left(m + \frac{I}{R_2^2}\right)}(1 - \cos\theta) \text{ (Eq. 2)}$$

Plug Eq. 2 into Eq. 1 to give:

$$\cos\theta = \frac{2m}{m + \frac{I}{R_2^2}}(1 - \cos\theta)$$

Solve for $\cos\theta$:

$$\cos\theta = \frac{2m}{3m + \frac{I}{R_2^2}}$$

For a sphere:

$$I = \frac{2}{5}mR_2^2$$

giving

$$\cos\theta = \frac{10}{17}$$

For a point of mass, $I = 0$, and $\cos\theta = \frac{2}{3}$.

Solution 2.04

Problem 2.04

This is a very common type of problem. The key is to write the energy in a form:

$$T = \frac{1}{2}m_{\text{eff}}\dot{x}^2$$

$$U = \frac{1}{2}k_{\text{eff}}x^2$$

Then the frequency of small oscillations is given by:

$$\omega = \sqrt{\frac{k_{\text{eff}}}{m_{\text{eff}}}}$$

In this particular case, $m_{\text{eff}} = m$. To get k_{eff}, write the total potential and then expand it in a Taylor series to quadratic order:

$$U = \frac{mM}{(d+x)^n} + \frac{mM}{(d-x)^n}$$

$$\frac{dU}{dx} = mM\left[\frac{-n}{(d+x)^{n+1}} + \frac{n}{(d+x)^{n+1}}\right] = 0 \text{ at } x = 0$$

$$\frac{d^2U}{dx^2} = mM\left[\frac{n(n+1)}{(d+x)^{n+2}} + \frac{n(n+1)}{(d-x)^{n+2}}\right] = \frac{2mMn(n+1)}{d^{n+2}} \text{ at } x = 0 = k_{\text{eff}}$$

Which gives:

29

$$\omega = \sqrt{\frac{2Mn(n+1)}{d^{n+1}}}$$

Solution 2.05

The orbit will be stable if the period of small oscillations about a perturbation from the orbit is finite.

$$F = cr^n$$

$$L = \frac{1}{2}m\dot{r}^2 + \frac{1}{2}mr^2\dot{\theta}^2 + \frac{c}{n+1}r^{n+1}$$

$$m\ddot{r} = \frac{l^2}{mr^3} + cr^n$$

where $l = mr^2\dot{\theta}$ is the angular momentum.

Now for a circle, $r = \rho$. Look at a small deviation from ρ:

$$\frac{l^2}{m\rho^3} = -c\rho^n \ (*)$$

$$\rho \rightarrow \rho + \Delta\rho$$

$$m\left(\ddot{\Delta\rho}\right) = \frac{l^2}{m(\rho + \Delta\rho)^3} + c(\rho + \Delta\rho)^n \approx \frac{-3l^2}{m\rho^4}\Delta\rho + nc\rho^{n-1}\Delta\rho$$

$$= \rho\left[nc\rho^{n-1} - \frac{3l^2}{m\rho^4}\right]$$

Substitute into $(*)$:

$$m\left(\ddot{\Delta\rho}\right) = (n+3)c\Delta\rho\rho^{n-1}$$

So k_{eff} is positive if $n + 3 > 0$, or $n > -3$.

Solution 2.06

Use the equations from Problem 2.05 to write:

$$F = cr^n$$

$$E = \frac{1}{2}m\dot{r}^2 + \frac{l^2}{mr^2} - \frac{c}{n+1}r^{n+1}$$

A circular orbit passing through the origin has the form:

$$r = 2r_0\cos\theta$$

$$\dot{r} = -r_0\sin\theta\dot{\theta} = \frac{-2r_0l}{mr^2}\sin\theta$$

which gives

$$E = \frac{1}{2}m\left(\frac{4r_0^2l^2}{m^2r^4}\sin^2\theta\right) + \frac{l^2}{mr^2} - \frac{c}{m+1}r^{n+1}$$

$$= \frac{l^2}{mr^2}\left[\frac{1}{2} + \frac{2r_0^2}{r^2}\sin^2\theta\right] - \frac{c}{n+1}r^{n+1} = \frac{l^2}{mr^2}\left[\frac{1}{2} + \frac{2r_0^2}{r^2}(1 - \cos^2\theta)\right] - \frac{c}{n+1}r^{n+1}$$

Substitute:

$$\cos\theta = \frac{r}{2r_0}$$

$$\Rightarrow E = \frac{l^2}{mr^2}\left[\frac{1}{2} + \frac{2r_0^2}{r^2}\left(1 - \frac{r^2}{4r_0^2}\right)\right] = -\frac{c}{n+1}r^{n+1} = \frac{2r_0^2l^2}{mr^4} - \frac{c}{n+1}r^{n+1}$$

So to have E constant, we must have $n+1 = -4$, or $n = -5$.

Solution 2.07

Problem 2.07

This is very straightforward; just want the kinetic energy to be equal to the gravitational potential energy:

$$\frac{1}{2}mv^2 = \frac{GmM}{r}$$

For the earth, the value is 11.2 km/s.

Solution 2.08

Problem 2.08

$$T = \frac{1}{2}ml^2\left(\dot{\theta}^2 + \dot{\varphi}^2\sin^2\theta\right)$$

$$L = \frac{1}{2}ml^2\left(\dot{\theta}^2 + \dot{\varphi}^2\sin^2\theta\right) + mgl\cos\theta$$

$$\frac{\delta L}{\delta \theta} = ml^2\sin\theta\cos\theta\dot{\varphi}^2 - mgl\sin\theta$$

$$\frac{\delta L}{\delta \dot{\theta}} = ml^2\dot{\theta}$$

$$\frac{\delta L}{\delta \varphi} = 0$$

$$\frac{\delta L}{\delta \dot{\varphi}} = ml^2\sin^2\theta\dot{\varphi}$$

This gives the equations of motion:

$$ml^2\ddot{\theta} = ml^2\sin\theta\cos\theta\dot{\varphi}^2 - mgl\sin\theta$$

$$\frac{d}{dt}\left(ml^2\sin^2\theta\dot{\varphi}\right) = 0$$

Solution 2.09

$$T = \frac{1}{2}m_1 \dot{x}^2 + \frac{1}{2}m_2 \dot{x}^2 + \frac{1}{2}I\omega^2$$

$$U = m_1 g x + m_2 g (C - x)$$

C is a constant; total length is constrained.

$$\omega = \frac{\dot{x}}{r}$$

$$\Rightarrow L = \frac{1}{2}\left(m_1 + m_2 = \frac{I}{r^2}\right)\dot{x}^2 + (m_2 - m_1)gx$$

$$\frac{d}{dt}\frac{\delta L}{\delta \dot{x}} - \frac{\delta L}{\delta x} = 0 = \left(m_1 + m_2 = \frac{I}{r^2}\right)\ddot{x} - (m_2 - m_1)g$$

Solution 2.10

$$T = \frac{1}{2}\vec{L} \cdot \vec{\omega} = \frac{1}{2}\sum_{\alpha\beta}\omega_\alpha I_{\alpha\beta}\omega_\beta$$

Maximum when? Look at expression about principal axes:

$$T = \frac{1}{2}\left(I_1\omega_1^2 + I_2\omega_2^2 + I_3\omega_3^2\right)$$

$$\omega^2 = \omega_1^2 + \omega_2^2 + \omega_3^2$$

$$\Rightarrow \dot{T} = \frac{1}{2}\left([I_1 + I_2 + I_3]\omega^2 - I_1[\omega_2^2 + \omega_3^2] - I_2[\omega_1^2 + \omega_3^2] - I_1[\omega_1^2 + \omega_2^2]\right)$$

If I_1 is the largest moment, then:

$$I_2 = I_1 - a$$

$$I_3 = I_1 - b$$

Maximize:

$$a\left(\omega_1^2 + \omega_3^2\right) + b\left(\omega_1^2 + \omega_2^2\right)$$

The minimum is when $\omega_2 = \omega_3 = 0$, or when the angular velocity corresponds with the axis with the largest moment.

Solution 2.11

Problem 2.11

The yo-yo is both rolling and falling. Initially have $U = mgh$; when the yo-yo has unrolled completely, we have:

$$U = mg\left(h - x\right) + \frac{1}{2}mv^2 + \frac{1}{2}I\omega^2$$

$$\omega = \frac{v}{r}$$

Set the initial and final energies equal and set $h = 0$ to give:

$$mgx = \frac{1}{2}v^2\left(m + \frac{I}{r^2}\right)$$

Differentiate with respect to time:

$$mgv = va\left(m + \frac{I}{r^2}\right) \Rightarrow a = \frac{mg}{m + \frac{I}{r^2}}$$

The tension is given by:

$$F = ma = mg - T \Rightarrow T = m\left(g - a\right) = \frac{mIg}{I + mr^2}$$

Solution 2.12

Problem 2.12

Conserve energy: $T = \frac{1}{2}mv^2$ and $U = 0$ both before and after the interaction. In the frame in which the center of mass velocity is 0, $v = v\hat{y} - V\hat{x}$, where V is the velocity of Jupiter. Then, the final velocity must be related to the initial velocity as:

$$\frac{1}{2}mv_f^2 = \frac{1}{2}mv^2$$

$$\left| v_f \right| = \sqrt{v^2 + V^2}$$

$$\vec{v_f} = \left| v_f \right| \cos\theta \hat{x} + \left| v_f \right| \sin\theta \hat{y}$$

$$= \left(\sqrt{v^2 + V^2}\cos\theta - V \right)\hat{x} + \sqrt{v^2 + V^2}\sin\theta \hat{y}$$

Solution 2.13

Problem 2.13

It sinks, because buoyancy effects disappear in free-fall.

Solution 2.14

Problem 2.14

a. The car will start slipping when the frictional force equals the gravitational force. For rear-wheel drive, the frictional force is operating to the rear of the center of mass; for front-wheel drive, it is in front of it. The equations are as follows.

Friction set equal to gravitation:

$2\mu N_{\text{rear}} = mg\sin\theta$ (for rear-wheel drive)

$2\mu N_{\text{front}} = mg\sin\theta$ (for front-wheel drive)

Normal forces:

$2N_{\text{rear}} + 2N_{\text{front}} = mg\cos\theta$

Moments about the pivot point:

$LN_{\text{front}} = N_{\text{rear}}\left(L - 2\mu h \right)$ (RWD)

$LN_{\text{rear}} = N_{\text{front}}\left(L + 2\mu h \right)$ (FWD)

Solve for the angle to give:

$$\tan\theta = \frac{\mu}{2 \pm \frac{2\mu h}{L}}$$

where the (+) is for front-wheel drive and the (-) is for rear.

b. The maximum acceleration is proportional to the friction. The acceleration can be written as a combination of static and dynamic parts:

$$F_{\text{front}} = \frac{mg}{2}\frac{L/2}{L} - \frac{mg}{2}\frac{h}{L}\frac{a}{g}$$

$$F_{\text{rear}} = \frac{mg}{2}\frac{L/2}{L} + \frac{mg}{2}\frac{h}{L}\frac{a}{g}$$

giving

$$ma_{\text{rwd}} = \mu mg\left(\frac{L/2}{L} + \frac{h}{L}\frac{a_{\text{rwd}}}{g}\right)$$

$$a_{\text{rwd}} = \frac{\mu g(L/2)}{L - \mu h}$$

Similarly, $a_{fwd} = \dfrac{\mu g(L/2)}{L + \mu h}$.

Solution 2.15

Problem 2.15

As in Electricity and Magnetism, we can use Gauss's law to calculate the gravitational potential at a radius r relative to a sphere of radius R_0:

$$\vec{g}\,(r) = \begin{bmatrix} -\frac{4}{3}\pi G\rho r\hat{r} \text{ for } r < R_0 \\ \hline -\frac{4}{3}\pi G\rho\frac{R_0^3}{r^2}\hat{r} \text{ for } r > R_0 \end{bmatrix}$$

$$\Phi(r) = -\int g(r)dr = \begin{bmatrix} \frac{2}{3}\pi G\rho\left(r^2 - 3R_0^2\right) \text{ for } r < R_0 \\ \\ -\frac{4}{3}\pi G\rho\rho\frac{R_0^3}{r^2} \text{ for } r > R_0 \end{bmatrix}$$

Now sum the potential as the superposition of the planet; the "hole" in the dust caused by the planet; and the outer ring of dust between the planet and the satellite to give:

$$\Phi(R) = -\frac{4}{3}\pi G\left[(\rho - \rho_0)\frac{R_0^3}{R} + \frac{\rho_0}{2}\left(3R_1^2 - R^2\right)\right], \text{ where the density of the planet is } \rho \text{ and that of}$$

the dust is ρ_0.

For circular orbits, the satellite is always at equilibrium distance:

$$m\omega^2 = \frac{1}{R}\frac{d\Phi}{dR} = -\frac{4}{3}\pi G\left[(\rho - \rho_0)\frac{R_0^3}{R} + \rho_0\right]$$

$$T = \frac{2\pi}{\omega} = \sqrt{\frac{3m}{4\pi G\left[(\rho - \rho_0)\frac{R_0^3}{R} + \rho_0\right]}}$$

For small oscillations about equilibrium, look at the orbital equation:

$$\ddot{r} - \frac{h^2}{r^3} = \frac{F(r)}{m}$$

where $h = r^2\dot{\theta}$. For small departures x from the equilibrium orbital radius R_{eq}:

$$\ddot{x} - \frac{h^2}{\left(R_{eq} + x\right)^3} = \frac{F\left(R_{eq} + x\right)}{m}$$

Expand to first order in x:

$$\ddot{x} - \frac{h^2}{R_{eq}^3}\left(1 - 3\frac{x}{R_{eq}}\right) = \frac{F\left(R_{eq}\right) + F'\left(R_{eq}\right)x}{m}$$

This gives harmonic motion with a period:

$$T = 2\pi \sqrt{\dfrac{m}{-3\dfrac{F\left(R_{\text{eq}}\right)}{R_{\text{eq}}} - F'\left(R_{\text{eq}}\right)}}$$

For our particular potential:

$$F = -\frac{d\Phi}{dr} = -\frac{4}{3}\pi G\left[(\rho - \rho_0)\frac{R_0^3}{R} + \rho_0 R\right]$$

$$\frac{dF}{dr} = -\frac{4}{3}\pi G\left[(-2)(\rho - \rho_0)\frac{R_0^3}{R} + \rho_0\right]$$

$$\Rightarrow -3F\left(R_{\text{eq}}\right) - F'\left(R_{\text{eq}}\right) = \frac{4}{3}\pi G\left[(\rho - \rho_0)\frac{R_0^3}{R_{\text{eq}}^3} + 4\rho_0\right]$$

$$T = 2\pi \sqrt{\dfrac{3m}{4\pi G\left[(\rho - \rho_0)\dfrac{R_0^3}{R_{\text{eq}}^3} + 4\rho_0\right]}}$$

Solution 2.16

Problem 2.16

$$L = \frac{m}{2}\left[A^2\omega^2\sin^2\omega t + l^2\dot{\theta}^2 - 2A\omega l\sin\omega t\,\dot{\theta}\cos\theta - 2gl(1 - \cos\theta)\right]$$

To get the behavior at small angle, write the equation of motion:

$$\frac{d}{dt}\left[l\dot{\theta} - A\omega\sin\omega t\cos\theta\right] = -g\sin\theta + A\omega\dot{\theta}\sin\omega t\sin\theta$$

$$\ddot{l\theta} - A\omega^2 \cos\omega t \cos\theta = -g\sin\theta$$

Now expand to first order in the small-angle approximation to give:

$$\ddot{l\theta} + g\theta = A\omega^2 \cos\omega t$$

This is just the equation for a driven harmonic oscillator with amplitude:

$$A = \frac{A\omega^2}{g - l\omega^2}$$

Solution 2.17

Problem 2.17

Using conservation of angular momentum would not be appropriate here, since this is a case of sudden impact with energy loss. The axles exert an external torque. Instead, we can relate the unknown applied force to the changes in angular velocity of the two gears (the primed quantities are those after the meshing):

$$-r_1 \int F dt = I_1(\omega_1{}' - \omega_1)$$

$$-r_2 \int F dt = I_2(\omega_2{}' - \omega_2)$$

$$\frac{r_2}{r_1} = \frac{I_2(\omega_2{}' - \omega_2)}{I_1(\omega_1{}' - \omega_1)}$$

Secondly, we can note that the velocity at the contact points has to be equal:

$$r_1\omega_1{}' = -r_2\omega_2{}'$$

which gives

$$r_1\omega_1{}' = -r_2\omega_2{}' = \frac{\left(\frac{I_1}{r_1}\right)\omega_1 - \left(\frac{I_2}{r_2}\right)\omega_2}{\frac{I_1}{r_1^2} + \frac{I_2}{r_2^2}}$$

Solution 2.18

$$L = \left(M + \frac{m}{2}\right)R^2\dot{\theta}^2 + \frac{m}{2}R^2\dot{\phi}^2 + mR^2\dot{\theta}\dot{\phi}\cos(\theta - \phi) + (M + m)gR\cos\theta + mgR\cos\phi$$

Now let:

$$\theta = \alpha e^{i\omega t}, \phi = \beta e^{i\omega t}$$

$\omega = \sqrt{\frac{g}{2R}}$ for the mode where $\alpha = \beta$

$\omega = \sqrt{\frac{m+M}{M}\frac{g}{R}}$ for the mode where $-(m + M)\alpha = m\beta$

Solution 2.19

The gravitational potential will vary according to the distance r from the center of the Earth:

$$F = ma = -\frac{GmM(r)}{r^2} = -\frac{Gm}{r^2}\frac{M_E}{\frac{4}{3}\pi R_E^3}\left(\frac{4}{3}\pi r^3\right) = -g\frac{r}{R_E}$$

Just a simple harmonic motion with:

$$T = 2\sqrt{\frac{R_E}{g}}$$

For a mass orbiting the Earth just above the surface, have to match the centripetal force to the force of gravity:

$$mg = \frac{mv^2}{R_E}$$

$$v = \sqrt{R_E g}$$

$$T = 2\sqrt{\frac{R_E}{g}}$$

So the period is the same as for the hypothetical mass through the earth.

Solution 2.20

$$\rho dV = \rho L r' dr' d\phi$$

$$d\vec{F} = \frac{Gm\rho}{r^2} dV$$

where $r^2 = r'^2 + R^2$

$$d\vec{F}_\parallel = dF\cos\theta = Gm\rho L \frac{Rr'}{\left(r'^2 + R^2\right)^{\frac{3}{2}}} dr' d\phi$$

$$\vec{F}_\parallel = 2\pi Gm\rho L \left[1 - \frac{z}{\left(z^2 + R^2\right)^{\frac{1}{2}}}\right] = m\ddot{z} = \frac{2GmM}{R^2}\left[1 - \frac{z}{\left(z^2 + R^2\right)^{\frac{1}{2}}}\right]$$

For small z:

$$F \approx \frac{2GmM}{R^2} = ma$$

$$d = \frac{1}{2}at^2$$

$$t = R\sqrt{\frac{d}{GM}}$$

Full period is $4t$; oscillations are non-harmonic.

Solution 2.21

$$m\frac{d^2x}{dt^2} + \mu\frac{dx}{dt} + 2k(x - x_0) = 0$$

$$x = x_0 + A\exp\left[\frac{-\mu + \sqrt{\mu^2 - 8km}}{2m}t\right] + B\exp\left[\frac{-\mu - \sqrt{\mu^2 - 8km}}{2m}t\right].$$

$$\omega = \sqrt{\frac{2k}{m} - \left(\frac{\mu}{2m}\right)^2}$$

Solution 2.22

a. Note that the moment of inertia about a fixed endpoint is

$$I_{end} = \frac{1}{12}ML^2 + M\left(\frac{L}{2}\right)^2.$$

For the upper pendulum portion, then:

$$K_{upper} = \frac{1}{2}I_{1,end}\dot{\theta}_1^2$$

$$V_{upper} = (2m)g\frac{D}{2}(1 - \cos\theta_1)$$

For the lower pendulum portion, consider things a bit differently. Take the center-of-mass motion:

$$K_{CM,lower} = \frac{1}{2}m\left[\left(\dot{\theta}_1(D)\cos\theta_1 + \dot{\theta}_2\left(\frac{2D}{2}\right)\cos\theta_2\right)^2 + \left(\dot{\theta}_1(D)\sin\theta_1 + \dot{\theta}_2\left(\frac{2D}{2}\right)\sin\theta_2\right)^2\right]$$

$$= \frac{1}{2}m\dot{\theta}_1^2D^2\cos^2\theta_1 + \dot{\theta}_2^2D^2\cos^2\theta_2 + 2\dot{\theta}_1\dot{\theta}_2D^2\cos\theta_1\cos\theta_2 + \dot{\theta}_1^2D^2\sin^2\theta_1 + \dot{\theta}_2^2D^2\sin^2\theta_2 + 2\dot{\theta}_1\dot{\theta}_2D^2\sin\theta_1\sin\theta_2$$

$$= \frac{1}{2}mD^2\left[\dot{\theta}_1^2 + \dot{\theta}_2^2 + 2\dot{\theta}_1\dot{\theta}_2(\cos\theta_1\sin\theta_2)\right]$$

$$= \frac{1}{2}mD^2\left[\dot{\theta}_1^2 + \dot{\theta}_2^2 + 2\dot{\theta}_1\dot{\theta}_2\cos(\theta_1 - \theta_2)\right]$$

$$K_{rot,lower} = \frac{1}{2} I_{2,CM} \dot{\theta_2}^2$$

And similarly with the potential:

$$V_{lower} = mg \Big(D(1 - \cos\theta_1) + 2D(1 - \cos\theta_2) \Big)$$

Now can write: $L = K - V$,

$$\frac{d}{dt} \left(\frac{\partial L}{\partial \dot{\theta_1}} \right) = \frac{\partial L}{\partial \theta_1}$$

And:

$$L = \frac{1}{2} I_{1,end} \dot{\theta_1}^2 + \frac{1}{2} mD^2 \left[\dot{\theta_1}^2 + \dot{\theta_2}^2 + 2\dot{\theta_1}\dot{\theta_2}\cos(\theta_1 - \theta_2) \right] + \frac{1}{2} I_{2,CM} \dot{\theta_2}^2$$

$$- (2m)g \frac{D}{2}(1 - \cos\theta_1) - mg \Big(D(1 - \cos\theta_1) + 2D(1 - \cos\theta_2) \Big)$$

Dropping constants

$$L = \frac{1}{2}(I_{1,end} + mD^2)\dot{\theta_1}^2 + \frac{1}{2}(I_{2,CM} + mD^2)\dot{\theta_2}^2 + mD^2\dot{\theta_1}\dot{\theta_2}\cos(\theta_1 - \theta_2) + 2mgD(\cos\theta_1 + \cos\theta_2)$$

Now expand the Lagrangian in the small-oscillation limit:

$$\cos(\theta_1 - \theta_2) \approx 1, \cos\theta \approx 1 - \frac{\theta^2}{2}$$

Then,

$$L \approx \frac{1}{2}(I_{1,end} + mD^2)\dot{\theta_1}^2 + \frac{1}{2}(I_{2,CM} + mD^2)\dot{\theta_2}^2 + mD^2\dot{\theta_1}\dot{\theta_2} + 2mgD\left(2 - \frac{\theta_1^2}{2} - \frac{\theta_2^2}{2}\right)$$

Now find the equations of motion based on this Lagrangian:

$$\frac{d}{dt} \left(\frac{\partial L}{\partial \dot{\theta_2}} \right) = \frac{\partial L}{\partial \theta_2}$$

$$\frac{d}{dt}\left((I_{2,CM} + mD^2)\dot{\theta}_2 + mD^2\dot{\theta}_1 \right) = -2mgD\theta_2$$

$$(I_{2,CM} + mD^2)\ddot{\theta}_2 + mD^2\ddot{\theta}_1 = -2mgD\theta_2$$

Further, using the other equation of motion:

$$\frac{d}{dt}\left(\frac{\partial L}{\partial \dot{\theta}_1} \right) = \frac{\partial L}{\partial \theta_1},$$

$$\frac{d}{dt}\left((I_{1,end} + mD^2)\dot{\theta}_1 + mD^2\dot{\theta}_2 \right) = -2mgD\theta_1$$

$$(I_{1,end} + mD^2)\ddot{\theta}_1 + mD^2\ddot{\theta}_2 = -2mgD\theta_1$$

Simplify a bit using:

$$I_{1,CM} + mD^2 = \frac{1}{12}(2m)D^2 + (2m)\frac{D^2}{4} + mD^2 = \frac{5}{3}mD^2$$

$$I_{2,CM} + mD^2 = \frac{1}{12}(m)(2D)^2 + mD^2 = \frac{4}{3}mD^2$$

Using these, see that:

$$\frac{4}{3}mD^2\ddot{\theta}_2 + mD^2\ddot{\theta}_1 = -2mgD\theta_2$$

$$\frac{5}{3}mD^2\ddot{\theta}_1 + mD^2\ddot{\theta}_2 = -2mgD\theta_1$$

$$\begin{bmatrix} \ddot{\theta}_1 \\ \ddot{\theta}_2 \end{bmatrix} = -2m\frac{g}{D}\begin{bmatrix} \left(\frac{5}{3} - \frac{3}{4}\right)^{-1} & \left(\frac{5}{3} - \frac{3}{4}\right)^{-1}\left(-\frac{3}{4}\right) \\ \left(\frac{4}{3} - \frac{3}{5}\right)^{-1}\left(-\frac{3}{5}\right) & \left(\frac{4}{3} - \frac{3}{5}\right)^{-1} \end{bmatrix}\begin{bmatrix} \theta_1 \\ \theta_2 \end{bmatrix}$$

The normal frequencies are then the eigenvalues of this matrix, in the form $\left| \sqrt{eigenvalue} \right|$ since the derivative is taken twice.

$$\begin{bmatrix} \ddot{\theta}_1 \\ \ddot{\theta}_2 \end{bmatrix} = 2m\frac{g}{D}\begin{bmatrix} \frac{12}{11} & \frac{12}{11}\left(-\frac{3}{4}\right) \\ \frac{15}{11}\left(-\frac{3}{5}\right) & \frac{15}{11} \end{bmatrix}\begin{bmatrix} \theta_1 \\ \theta_2 \end{bmatrix} = -\frac{2}{11}m\frac{g}{D}\begin{bmatrix} 12 & -9 \\ -9 & 15 \end{bmatrix}\begin{bmatrix} \theta_1 \\ \theta_2 \end{bmatrix}$$

$$C = -\frac{2}{11}m\frac{g}{D}$$

$$(12C - \lambda)(15C - \lambda) - 81C^2 = 0$$

$$\lambda^2 - 27C\lambda + 99C^2 = 0$$

$$\lambda_\pm = \left(-\frac{2}{11}m\frac{g}{D}\right)\frac{27 \pm \sqrt{27^2 - 4 \cdot 99}}{2} = \left(-\frac{6}{11}m\frac{g}{D}\right)\frac{9 \pm \sqrt{37}}{2}$$

$$\omega_{1,2} = \left| \sqrt{\lambda_{-,+}} \right|$$

In this case, ω_1 will be the minus combination and ω_2 will be the plus combination in order to minimize and maximize the respective coefficients.

b. These coordinates then correspond to the eigenvectors that go with these eigenvalues. The easiest way to find these will be to simply find the eigenvectors of $\begin{bmatrix} 12 & -9 \\ -9 & 15 \end{bmatrix}$, or $\begin{bmatrix} 4 & -3 \\ -3 & 5 \end{bmatrix}$, which will be the same eigenvectors within a constant factor. In order to find them, take:

$$\begin{bmatrix} 4 & -3 \\ -3 & 5 \end{bmatrix} \rightarrow (4 - \lambda)(5 - \lambda) - 9 = 0 \rightarrow \lambda^2 - 9\lambda + 11 = 0 \rightarrow \lambda_\pm = \frac{9 \pm \sqrt{81 - 44}}{2} = \frac{9 \pm \sqrt{37}}{2}$$

And now the eigenvectors are within a constant factor the null spaces of

$$\begin{bmatrix} 4 - \lambda_+ & -3 \\ -3 & 5 - \lambda_+ \end{bmatrix} \text{ and } \begin{bmatrix} 4 - \lambda_- & -3 \\ -3 & 5 - \lambda_- \end{bmatrix} \text{ with implicit multiples of } \begin{bmatrix} \theta_1 \\ \theta_2 \end{bmatrix}. \text{ These occur where}$$

$$(4 - \lambda_+)c_1 - 3c_2 = 0$$

$$-3c_1 + (5 - \lambda_+)c_2 = 0$$

and

$$(4 - \lambda_-)c_1 - 3c_2 = 0$$
$$-3c_1 + (5 - \lambda_-)c_2 = 0,$$ respectively. Thus,

$$c_1 = \frac{3}{(4 - \lambda_\pm)}c_2, \text{ and so } \eta_\pm = \theta_1 + \frac{3}{(4 - \lambda_\pm)}\theta_2 \text{ to within a constant factor.}$$

Solution 2.23

Problem 2.23

a. Assume that k is negative.

$$V_g = mgz$$

$$V_c = \int_0^{\sqrt{p^2 + z^2}} krdr = \frac{1}{2}k(p^2 + z^2) = \frac{1}{2}k(R^2 + z^2)$$

$$V = V_g + V_c$$

$$K = \frac{1}{2}m\left(\dot{z}^2 + (\rho\dot{\phi})^2 + \dot{\rho}^2\right) = \frac{1}{2}m\left(\dot{z}^2 + (R\dot{\phi})^2\right)$$

b. $L = K - V$

$$= \frac{1}{2}m\left(\dot{z}^2 + (\rho\dot{\phi})^2 + \dot{\rho}^2\right) - mgz - \frac{1}{2}k(\rho^2 + z^2)$$

$$\frac{d}{dt}\left(\frac{\partial L}{\partial \dot{z}}\right) = \frac{\partial L}{\partial z}$$

$$\frac{d}{dt}(m\dot{z}) = -mg - kz$$

$$m\ddot{z} = -mg - kz$$

$$\frac{d}{dt}\left(\frac{\partial L}{\partial \dot{\theta}}\right) = \frac{\partial L}{\partial \theta}$$

$$\frac{d}{dt}\left(m\rho^2\dot{\phi}\right) = 0$$

$$\vec{F}_{cyl} = \hat{\rho}\frac{\partial L}{\partial \rho} = \hat{\rho}\left(m\rho\dot{\phi}^2 - k\rho\right)$$

This force is the familiar central potential plus the centripetal acceleration.

c. The magnitude of the force is given by

$$\vec{F}_{cyl} \cdot \hat{\rho} = F_{cyl} = \left(mR\dot{\phi}^2 - kR\right)$$

d. From the second equation of motion, it is clear that angular momentum is conserved: this is certainly always the case, and, in this case, the angular momentum of the particle will always remain such that no force ever acts in the rotational direction. Further, energy of the particle is conserved as this is a frictionless scenario. The "normal" force acting perpendicular to the allowed motion of the particle can do no work.

e. Certainly, $\rho = R$ as the particle remains on the surface of the cylinder. Since angular momentum is conserved, $\dfrac{d}{dt}\left(mR^2\dot{\phi}\right) = 0$, $\dot{\phi} = const$, $\phi(t) = \dot{\phi}_0 t$. All that remains is to solve the differential equation in z. Have $m\ddot{z} = -mg - kz$; the solution to the homogenous equation is clearly of the form $m\ddot{z} = -kz$, $z \approx A\cos\sqrt{\dfrac{k}{m}}t + B\sin\sqrt{\dfrac{k}{m}}t$. A particular solution is of the form $z = -\dfrac{mg}{k}$. Solutions to the full equation are then

$z \approx A\cos\sqrt{\dfrac{k}{m}}t + \sin\sqrt{\dfrac{k}{m}}t - \dfrac{mg}{k}$. Using boundary conditions:

$$0 = A\cos\sqrt{\tfrac{k}{m}}0 + B\sin\sqrt{\tfrac{k}{m}}0 - \tfrac{mg}{k} = A - \tfrac{mg}{k}$$

$$0 = -A\sin\sqrt{\tfrac{k}{m}}0 + B\cos\sqrt{\tfrac{k}{m}}0 = B$$

$$z(t) = \tfrac{mg}{k}\cos\sqrt{\tfrac{k}{m}}t - \tfrac{mg}{k}$$

$$\rho(t) = R$$

$$\phi(t) = \dot{\phi}_0 t$$

Solution 2.24

Problem 2.24

a. First, need to know $\rho(0)$ in terms of ball mass m and radius a:

$$m = 4\pi \int_0^a \rho(r)r^2 dr = 4\pi\rho(0)\int_0^a \left[1 - \tfrac{4r}{5a}\right]r^2 dr = 4\pi\rho(0)\left[\tfrac{1}{3}a^3 - \tfrac{4}{5a}\tfrac{1}{4}a^4\right]$$

$$\rho(0) = \frac{m}{4\pi\left[\tfrac{1}{3}a^3 - \tfrac{1}{5}a^3\right]} = \frac{15m}{8\pi a^3}$$

Certainly,

$$I = \int_M r_{axis}^2 dm = \int_M r_{axis}^2 \frac{dm}{dx^3} dx^3 = \int_M [r\sin\phi]^2 \rho(0)\left[1 - \frac{4r}{5a}\right] r^2 \sin\phi \, dr \, d\theta \, d\phi$$

$$= \rho(0)\int_0^a \int_0^{2\pi} \int_0^{\pi} \left[r^4 - \frac{4r^5}{5a}\right]\sin^3\phi \, dr \, d\theta \, d\phi = 2\pi\rho(0)\left[\frac{1}{5}a^5 - \frac{1}{6}\frac{4a^6}{5a}\right]\int_0^{\pi} \sin^3\phi \, d\phi$$

In order to complete this integral, take:

$$\int_0^{\pi} \sin^3\phi \, d\phi = \int_0^{\pi} (1 - \cos^2\phi)\sin\phi \, d\phi = \int_0^{\pi} \sin\phi \, d\phi - \int_0^{\pi} \sin\phi\cos^2\phi \, d\phi$$

$$= 2 - \int_{\pi}^{0} \sin\phi\cos^2\phi \, d\phi$$

$$u = \cos\phi, \, du = -\sin\phi$$

$$= 2 + \int_{-1}^{1} u^2 du = 2 + \frac{1}{3}u^3 \Big|_{-1}^{1} = \frac{4}{3}$$

Collecting these results, have:

$$I = 2\pi \frac{15m}{8\pi a^3}\left[\frac{1}{5}a^5 - \frac{1}{6}\cdot\frac{4a^6}{5a}\right]\frac{4}{3} = \frac{m}{a^3}\left[a^5 - \frac{2}{3}a^5\right] = \frac{1}{3}ma^2$$

b. For the purposes of tracking the different values through this problem, let
$R = 4a, M = 5m, I = \frac{1}{3}ma^2$, and of course $L = K - V$.

There are four terms to consider here:

The kinetic energy of the center-of-mass motion of the wedge.

$$K_{CMwedge} = \frac{1}{2}M\dot{x}^2$$

48

The kinetic energy of the center-of-mass motion of the ball.

$$K_{CMball} = \frac{1}{2}m\left[\left(\dot{x} + (R-a)\dot{\theta}\sin\theta\right)^2 + \left((R-a)\dot{\theta}\cos\theta\right)^2\right]$$

$$K_{CMball} = \frac{1}{2}m\left[\dot{x}^2 + 2(R-a)\dot{\theta}\dot{x}\sin\theta + (R-a)^2\dot{\theta}^2\right]$$

The rotational kinetic energy of the ball.

Again, consider the rotational kinetic energy of the ball: recall the no-slipping condition.

$$K_{Rotball} = \frac{1}{2}I\omega^2$$

$$\omega = \left(\frac{R}{a}\dot{\theta} - \dot{\theta}\right)$$

$$K_{Rotball} = \frac{1}{2}I\left(\frac{R}{a}\dot{\theta} - \dot{\theta}\right)^2$$

Why does this discrepancy exist? Here is the fallacy that one might fall into: There is always a one-to-one correspondence between points on the ball and points on the surface of the wedge such that an equal distance is traversed by the contact point about the outside of the ball as the length of the surface. However, it is a fallacy to believe that this length is then equal to the actual rotation that has occurred with respect to a Cartesian coordinate system. We must correct for this precession about the outside of the ball by subtracting off $\dot{\theta}$ in the central coordinate system, as this point will undoubtedly precess exactly once per rotation about the large circle. (Imagine an enormous inside circle rolling inside a slightly larger outer circle. Will it rotate very nearly once as it rolls about the inside of the outer circle? No, in fact it rotates very slowly).

The gravitational potential energy of the ball.

Assume that the scale of this picture is small enough such that the acceleration due to gravity is constant.

$$V_{ball} = (R-a)mg\left(1 - \sin\theta\right)$$

$$L = \frac{1}{2}(M+m)\dot{x}^2 + \frac{1}{2}\left(m(R-a)^2 + I\left(\frac{R}{a} - 1\right)^2\right)\dot{\theta}^2 + m(R-a)\dot{\theta}\dot{x}\sin\theta - (R-a)mg\left(1 - \sin\theta\right)$$

$$= 3\dot{x}^2 + 6ma^2\dot{\theta}^2 + 3am\dot{\theta}\dot{x}\sin\theta - 3amg\left(1 - \sin\theta\right)$$

c. Using $\frac{d}{dt}\left(\frac{\partial L}{\partial \dot{x}}\right) = \left(\frac{\partial L}{\partial x}\right)$ gives

$\frac{d}{dt}\left((M+m)\dot{x} + m(R-a)\dot{\theta}\sin\theta\right) = 0$

$(M+m)\ddot{x} + m(R-a)\left(\ddot{\theta}\sin\theta + \dot{\theta}^2\cos\theta\right) = 0$

Further:

$\frac{d}{dt}\left(\frac{\partial L}{\partial \dot{\theta}}\right) = \left(\frac{\partial L}{\partial \theta}\right)$

$\frac{d}{dt}\left(\left(m(R-a)^2 + I\left(\frac{R}{a}-1\right)^2\right)\dot{\theta} + m(R-a)\dot{x}\sin\theta\right) = m(R-a)\cos\theta(\dot{\theta}\dot{x}+g)$

$\left(m(R-a)^2 + I\left(\frac{R}{a}-1\right)^2\right)\ddot{\theta} + m(R-a)\left(\ddot{x}\sin\theta + \dot{x}\dot{\theta}\cos\theta\right) = m(R-a)\cos\theta(\dot{\theta}\dot{x}+g)$

Simplifying:

$\ddot{x} = -\frac{m(R-a)}{(M+m)}\left(\ddot{\theta}\sin\theta + \dot{\theta}^2\cos\theta\right)$

$\ddot{x} = \frac{\cos\theta}{\sin\theta}(\dot{\theta}\dot{x}+g) - \frac{1}{\sin\theta}\left((R-a) + \frac{1}{m(R-a)}\left(\frac{R}{a}-1\right)^2\right)\ddot{\theta} - \dot{x}\dot{\theta}\frac{\cos\theta}{\sin\theta}$

For:

$\frac{\cos\theta}{\sin\theta}(\dot{\theta}\dot{x}+g) - \frac{1}{\sin\theta}\left((R-a) + \frac{I}{m(R-a)}\left(\frac{R}{a}-1\right)^2\right)\ddot{\theta} - \dot{x}\dot{\theta}\frac{\cos\theta}{\sin\theta} = -\frac{m(R-a)}{(M+m)}\left(\ddot{\theta}\sin\theta + \dot{\theta}^2\cos\theta\right)$

$\cos\theta(\dot{\theta}\dot{x}+g) - \left((R-a) + \frac{I}{m(R-a)}\left(\frac{R}{a}-1\right)^2\right)\ddot{\theta} - \dot{x}\dot{\theta}\cos\theta = -\sin\theta\frac{m(R-a)}{(M+m)}\left(\ddot{\theta}\sin\theta + \dot{\theta}^2\cos\theta\right)$

$\left(\sin^2\theta\frac{m(R-a)}{(M+m)} - (R-a) - \frac{I}{m(R-a)}\left(\frac{R}{a}-1\right)^2\right)\ddot{\theta} = -\sin\theta\frac{m(R-a)}{(M+m)}\left(\dot{\theta}^2\cos\theta\right) - \cos\theta(\dot{\theta}\dot{x}+g) + \dot{x}\dot{\theta}\cos\theta$

$\left(\sin^2\theta\frac{m(R-a)}{(M+m)} - (R-a) - \frac{I}{m(R-a)}\left(\frac{R}{a}-1\right)^2\right)\ddot{\theta} = -\sin\theta\frac{m(R-a)}{(M+m)}\left(\dot{\theta}^2\cos\theta\right) - g\cos\theta$

$$\ddot{\theta} = \frac{\sin\theta \frac{m(R-a)}{(M+m)}\left(\dot{\theta}^2\cos\theta\right) + g\cos\theta}{(R-a) + \frac{I}{m(R-a)}\left(\frac{R}{a}-1\right)^2 - \sin^2\theta\frac{m(R-a)}{(M+m)}} \rightarrow \left[\frac{\dot{\theta}^2\sin\theta + \frac{2g}{a}}{8 - \sin^2\theta}\right]\cos\theta$$

Now have:

$$\ddot{x} = -\frac{m(R-a)}{(M+m)}\left(\frac{\sin\theta\frac{m(R-a)}{(M+m)}\left(\dot{\theta}^2\cos\theta\right) + g\cos\theta}{(R-a) + \frac{I}{m(R-a)}\left(\frac{R}{a}-1\right)^2 - \sin^2\theta\frac{m(R-a)}{(M+m)}}\sin\theta + \dot{\theta}^2\cos\theta\right)$$

$$\rightarrow -\frac{a}{2}\left(\left[\frac{\dot{\theta}^2\sin\theta + \frac{2g}{a}}{8 - \sin^2\theta}\right]\cos\theta\sin\theta + \dot{\theta}^2\cos\theta\right)$$

d. Solving these differential equations seems unnecessarily complex. Instead, let's use a time-honored bit of treachery to determine the solution. Look at the total kinetic and potential energies: certainly, the ball has traveled a certain height downward at this time:

$$K + V = const$$

$$\tfrac{1}{2}(M+m)\dot{x}^2 + \tfrac{1}{2}\left(m(R-a)^2 + I\left(\frac{R}{a}-1\right)^2\right)\dot{\theta}^2 + m(R-a)\dot{\theta}\dot{x}\sin\theta = K$$

$$(R-a)mg\left(1 - \sin\theta\right) = V$$

At the bottom of the slope, when the potential energy has become kinetic, this becomes

$$\frac{1}{2}(M+m)\dot{x}^2 + \frac{1}{2}\left(m(R-a)^2 + I\left(\frac{R}{a}-1\right)^2\right)\dot{\theta}^2 + m(R-a)\dot{\theta}\dot{x} = (R-a)mg$$

or

$$3m\dot{x}^2 + \tfrac{1}{2}\left(m(3a^2) + \tfrac{1}{3}ma^2 9^2\right)\dot{\theta}^2 + m(3a)\dot{x}\dot{\theta} = 3amg$$

$$3\dot{x}^2 + 6a^2\dot{\theta}^2 + 3a\dot{\theta}\dot{x} = 3ag$$

Further, know that net momentum in the x-direction is conserved:

$$M\dot{x} + m\left(\dot{x}(R-a)\dot{\theta}\sin\theta\right) = 0$$

$$\dot{\theta} = -\frac{(M+m)}{m(R-a)\sin\theta}\dot{x} \rightarrow -\frac{2}{a}\dot{x}$$

Overall, then, have:

$$3\dot{x}^2 + 6a^2\left(\frac{2}{a}\right)^2 \dot{x}^2 - 3a\frac{2}{a}\dot{x}^2 = 3ag$$

$$(3 + 24 - 6)\dot{x}^2 = 3ag$$

$$\dot{x} = -\sqrt{\frac{ag}{7}}$$

Choose the negative root since the wedge will clearly end up going to the left. Note that the ball will start slipping after it leaves the wedge: the angular velocity required to roll without slipping is discontinuous here.

Solution 2.25

a. The terms to consider are:

The kinetic energy of the main body of the sled, $K_{body} = \frac{1}{2}\beta\dot{s}^2$;

the kinetic energy of the pendulum bob,

$$K_{bob} = \frac{1}{2}m\left[\left(b\dot{\phi}\cos(\phi + \alpha) + \dot{s}\right)^2 + \left(b\dot{\phi}\sin(\phi + \alpha)\right)^2\right] = \frac{1}{2}m\left[2b\dot{\phi}\dot{s}\cos(\phi + \alpha) + \dot{s}^2 + \left(b\dot{\phi}\right)^2\right];$$

the potential energy of the main body of the sled, $V_{body} = \beta mg(-s\sin\alpha) + C_{1,geometry}$, where the constant does not factor into any equation of motion, it simply represents that this "center of mass" potential does not take into account any rearrangement of the orientation of the sled;

the potential energy of the pendulum bob, $V_{bob} = mg(-s\sin\alpha) + mgb(1 - \cos\phi) + C_{2,geometry}$,

where again the constant represents some constancy of geometry like the height of the mast from which the bob hangs and does not factor into any equation of motion.

The relevant part of the Lagrangian, then, is

$$L = K - V = \frac{1}{2}m\beta m\dot{s}^2 + \frac{1}{2}m\left(2b\dot{\phi}\dot{s}\cos(\phi + \alpha) + \dot{s}^2 + \left(b\dot{\phi}\right)^2\right) - \beta mg(-s\sin\alpha) - mg(-s\sin\alpha) - mgb(1 - \cos\phi)$$

b.

$$p_s = \frac{\partial L}{\partial \dot{s}}$$

$$p_s = \beta m\dot{s} + \frac{1}{2}m\left(2b\dot{\phi}\cos(\phi + \alpha) + 2\dot{s}\right)$$

$$p_\phi = \frac{\partial L}{\partial \dot{\phi}}$$

$$p_\phi = \frac{1}{2}m\left(2b\dot{s}\cos(\phi + \alpha) + 2b^2\dot{\phi}\right)$$

c. The Hamiltonian is given by $H = \sum_i \dot{q}_i p_i - L = \dot{s}p_s + \dot{\phi}p_\phi - L$. The only step necessary to obtain the form $H\left(\phi, s, p_\phi, p_s\right)$, then, is to solve the expressions for conjugate momenta in terms of the time derivatives of the coordinates and substitute these into the Lagrangian above.

d.

$$\frac{d}{dt}\left(\frac{\partial L}{\partial \dot{s}}\right) = \frac{\partial L}{\partial s}$$

$$\frac{d}{dt}\left[\beta m\dot{s} + \frac{1}{2}m\left(2b\dot{\phi}\cos(\phi + \alpha) + 2\dot{s}\right)\right] = -\beta mg(-\sin\alpha) - mg(-\sin\alpha)$$

$$\left(\beta m\ddot{s} + \frac{1}{2}m\left(2b\ddot{\phi}\cos(\phi + \alpha) - 2b\dot{\phi}^2\sin(\phi + \alpha) + 2\ddot{s}\right)\right) = \beta mg(\sin\alpha) + mg(\sin\alpha)$$

$$\left[(\beta + 1)m\ddot{s} + mb\left(\ddot{\phi}\cos(\phi + \alpha) - \dot{\phi}^2\sin(\phi + \alpha)\right)\right] = (\beta + 1)mg(\sin\alpha)$$

$$\frac{d}{dt}\left(\frac{\partial L}{\partial \dot{\phi}}\right) = \frac{\partial L}{\partial \phi}$$

$$\frac{d}{dt}\left(\frac{1}{2}m\left(2b\dot{s}\cos(\phi + \alpha)2b^2\dot{\phi}\right)\right) = -\frac{1}{2}m\left(2b\dot{\phi}\dot{s}\sin(\phi + \alpha)\right) - mgb\sin\phi$$

$$\frac{1}{2}m\left(2b\ddot{s}\cos(\phi + \alpha) - 2b\dot{s}\dot{\phi}\sin(\phi + \alpha) + 2b^2\ddot{\phi}\right) = -\frac{1}{2}m\left(2b\dot{\phi}\dot{s}\sin(\phi + \alpha)\right) - mgb\sin\phi$$

$$mb\left(\ddot{s}\cos(\phi + \alpha) + b\ddot{\phi}\right) = -mgb\sin\phi$$

e. Take the above equations of motion with $\phi \to \phi_0$, $\dot{\phi} \to 0$, $\ddot{\phi} \to 0$.

53

$$(\beta + 1)m\ddot{s} = (\beta + 1)mg(\sin\alpha)$$

$$mb\ddot{s}\cos(\phi + \alpha) = -mgb\sin\phi_0$$

and

$$\ddot{s} = g(\sin\alpha)$$

$$\ddot{s}\cos(\phi + \alpha) = -g\sin\phi_0$$

$$\sin\alpha\cos(\phi + \alpha) + \sin\phi_0 = 0$$

References
Classical mechanics

1. Bayman, B.F. & Hamermesh, M. *A Review of Undergraduate Physics*, (Wiley, New York, 1986).
2. Goldstein, H., Poole, C.P. & Safko, J.L. *Classical Mechanics*, (Addison Wesley, San Francisco, 2002).
3. Marion, J.B. & Thornton, S.T. *Classical Dynamics of Particles & Systems*, (Harcourt Brace Jovanovich, San Diego, 1988).
4. Fetter, A.L. & Walecka, J.D. *Theoretical Mechanics of Particles and Continua*, (Dover Publications, Mineola, N.Y., 2003).

3

Special relativity

Problems
Special relativity

Problem 3.01   Solution 3.01

Derive the Compton equation relating the wavelength of the incident photon λ_i and scattered photon λ_f with the mass of the electron and scattering angle θ:

$$\lambda_f - \lambda_i = \frac{1 - \cos\theta}{m_e c^2}$$

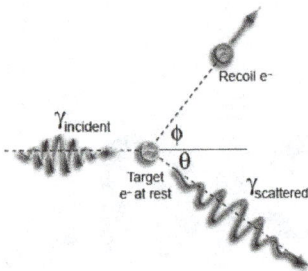

Problem 3.02   Solution 3.02

A Compton scattering experiment is set up so that the scattered photon and recoil electron are detected only if they are at right angles to each other. What is the energy of the scattered photon?

Problem 3.03   Solution 3.03

In a Compton scattering experiment, what is the maximum scattering angle at which the scattered photon is capable of producing an electron-positron pair?

Problem 3.04

Solution 3.04

Find the accelerating voltage required for an electron so that the phase velocity of the wave is twice the electron's velocity.

Problem 3.05

Solution 3.05

For experimental detection of the antiproton, a collider was designed in the 1950's to look at the reaction $p + p \rightarrow p + p + p + \bar{p}$, where a stationary proton is collided with a fast-moving proton.

a. What is the threshold energy for this high-energy proton?

b. Is the reaction $p + p \rightarrow p + p + \pi^+ + \bar{p}$ possible? If not, why?

Problem 3.06

Solution 3.06

A mass of m decays at rest into a proton and a photon. If the photon has a wavelength of $\lambda = 1$ fm (10^{-15} m) what is m? (The proton mass is $m_p = 938 \; \frac{\text{MeV}}{c^2}$.)

Problem 3.07

Solution 3.07

The $n = 3$ to $n = 2$ hydrogen line (hydrogen-alpha) from a distant galaxy is observed to be red-shifted by an amount of $\Delta\lambda = 100$ nm. How fast is the galaxy moving away from us?

Problem 3.08

Solution 3.08

In a famous experiment, a detector was placed on Mt. Washington (about 2,000 meters above sea level) to measure muon flux. By how much would the flux be reduced if the particles traveled to sea level? Consider only decay of the particles, which have a velocity of $0.98c$ when produced and a half-life of 1.56 μs.

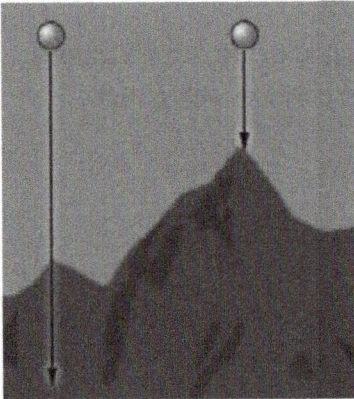

Problem 3.09

Solution 3.09

Measurements of planetary magnetic fields by the Charge Composition Explorer spacecraft have shown no fall-off of the electromagnetic force over 5×10^{-8} m. What upper bound does this place on the mass of the photon?

Problem 3.10

Solution 3.10

Another way to put a limit on the photon mass was suggested by de Broglie in 1940. He suggested that light from a distant star would arrive at Earth as a function of wavelength. For a star 10^3 light-years distant, and a possible time resolution of the detector of 1 ms, what is the upper limit on the mass of the photon? Assume the detector spans the visible spectrum — that is, the biggest difference is seen between a red photon and a blue one.

Problem 3.11

Solution 3.11

A photon and a nucleus (mass m) are traveling toward each other at relativistic speeds. If the energy of the photon in the lab frame is E_0, and the energy of the nucleus's first excited state is also E_0, what velocity must the nucleus have in order to absorb the photon? Where in the EM spectrum would you expect this photon to be found?

Problem 3.12

Solution 3.12

An astronaut travels to the outer planets at a speed of $0.9c$. In his time frame, the trip takes a week.

a. How long does the trip take according to an observer on Earth?

b. What is the distance traveled in the astronaut's reference frame?

c. At what speed is the target planet approaching in the astronaut's frame of reference?

Problem 3.13

Solution 3.13

In Mossbauer experiments, ^{191}Ir emits a 129 keV gamma ray line with a width at half-maximum of 4.6×10^{-6} eV.

a. What is the mean lifetime of the excited state that leads to this line?

b. What relative velocity of source and observer would lead to a Doppler shift of this line equal to the linewidth?

Problem 3.14

Solution 3.14

The universe is filled with thermal photons left over from the Big Bang, with energy kT where $T \sim 3$ K. High-energy protons traveling through space can collide with these photons to produce a neutron and positively-charged pion. What is the minimum proton momentum required for this to occur?

Problem 3.15

Solution 3.15

What is the shift in the hydrogen atom ground state due to relativistic effects on the electron mass?

Problem 3.16

Solution 3.16

A rod of length L making an angle of θ with respect to the x-axis in its own frame, called x', moves along that axis with velocity v. What is the length as seen by a stationary observer? What is the angle it makes with the observer's x-axis?

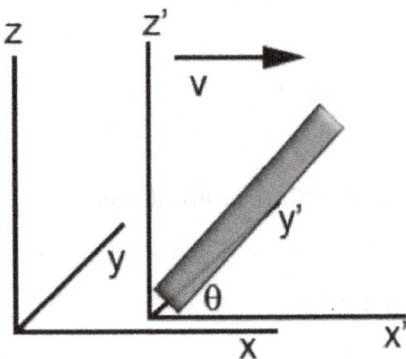

Problem 3.17

Solution 3.17

This problem concerns the precession of perihelion predicted by general relativity and how it might be generated by modifications of the distribution of mass in the sun.

a. Write down the Lagrangian for the central force problem for a particle restricted to the equatorial plane $\left(\theta = \dfrac{\pi}{2} \right)$ and show that the Euler-Lagrange equations are:

$$m\ddot{r} = mr\dot{\phi}^2 - \frac{\partial V(r)}{\partial r}$$

$$\frac{d}{dt}\left(mr^2\dot{\phi} \right) = 0$$

where r is the magnitude of the radius vector connecting the sun to the planet (still in equatorial plane), ϕ is the angle measured from the perihelion or closest approach of the planet to the sun and $V(r)$ is an arbitrary central potential. What is the physical interpretation of the second equation?

b. Consider a thin uniform ring of matter of radius h and mass M_{ring} as shown in the figure.

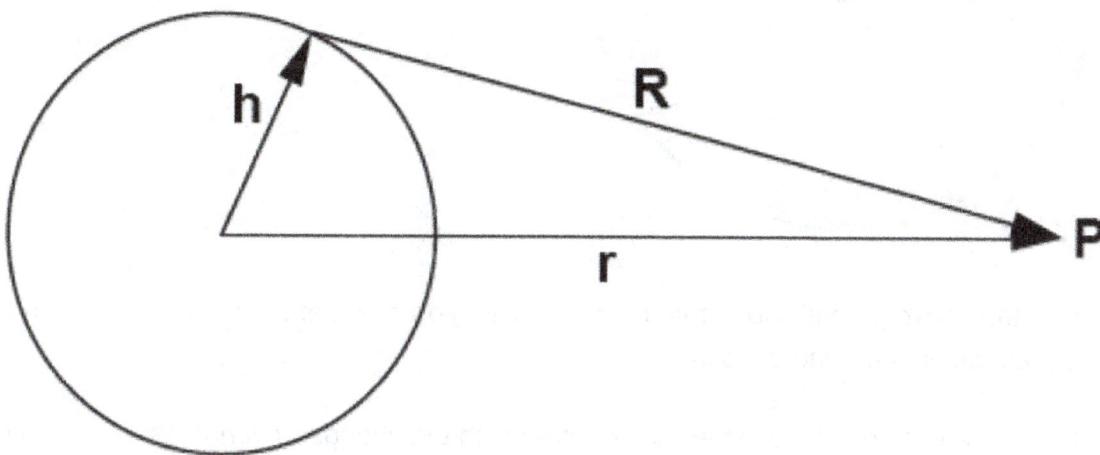

Write down an integral expression, including limits of integration, for the gravitational potential energy per unit mass at the point P in the figure. Note that P lies in the plane of the ring. Do not compute the integral at this time.

c. Expand the integrand that you found above in the limit that $r \gg h$, keeping terms up to order $\dfrac{h^2}{r^2}$. Integrate this expression to find the potential at P and check that you have arrived at the expected answer in the limit $r \to \infty$.

d. Suppose we model a deviation from a spherically symmetric mass distribution for the Sun by adding a "bulge" to it given by a thin ring of mass as above at the Sun's equator.

Take the mass and radius of the sun to be M_{sun}, R_{sun}, while the mass of the bulge will be M_{bulge} and its radius will be that of the sun. Add the gravitational potentials from both the spherical mass distribution as well as that of the bulge to find the gravitational potential energy of a mass M at a point P. The distance r is described in part (b).

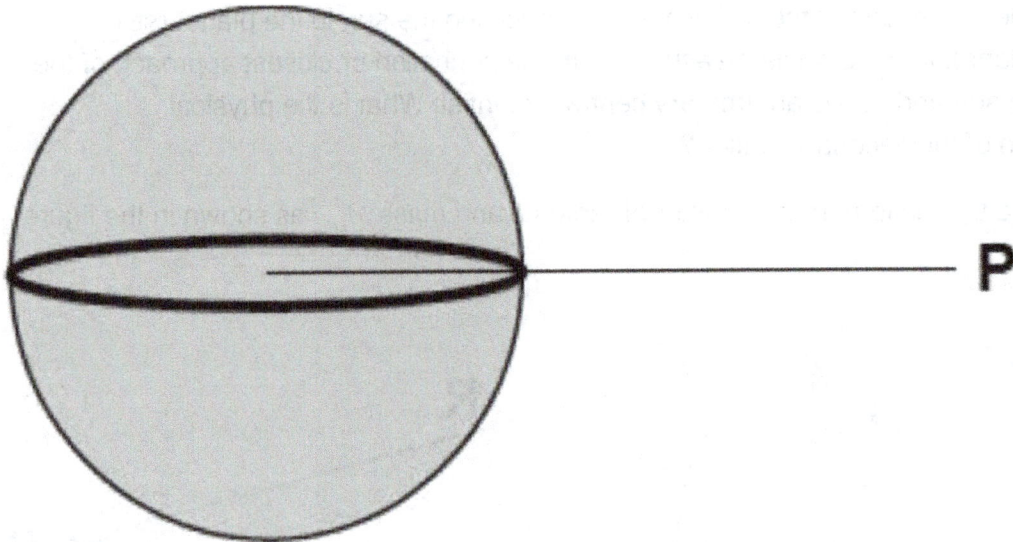

e. Use the above potential to derive the force acting on a planet of mass m moving in equatorial orbit around the sun.

f. Use the equations of motion above together with the change of variables $u = \dfrac{1}{r}$ to find that for a general central potential $V(r)$ we can write

$$\frac{d^2 u(\phi)}{d\phi^2} + u(\phi) = -\frac{m}{u^2 L^2} F(u)$$

where we have changed the independent variable from t to the angle ϕ. $F(u)$ is the force $F(r)$ derived from the potential $V(r)$ after the substitution $u = \frac{1}{r}$, and L is the constant angular momentum about the axis perpendicular to the plane of the orbit. Apply this formula to the potential found for the sun together with the bulge.

g. The force law found by adding the bulge to the mass distribution of the sun is of the same form as the modification given by Einstein's theory of general relativity. The full equation of motion can be written as:

$$\frac{d^2u}{d\phi^2} + u = \frac{1}{\alpha} + \delta u^2,$$

where δ is constant. We consider the limit in which $\frac{\delta}{\alpha} \ll 1$. Solve this equation in the limit that the u^2 term can be neglected. Describe a procedure that you might use to get an approximate solution to the full equation. How could you use this to get the precession of perihelion due to the bulge?

Solutions
Special relativity

Solution 3.01

Problem 3.01

Conservation of momentum:

$$\vec{p}_i = \vec{p}_f + \vec{p}_e$$

$$p_e^2 = \left(\vec{p}_i - \vec{p}_f\right)\left(\vec{p}_i - \vec{p}_f\right) = p_i^2 + p_f^2 - 2p_ip_f\cos\theta$$

Conservation of energy:

$$E_i + m_ec^2 = E_f + \sqrt{m_e^2c^4 + p_f^2c^2}$$

Equate the two expressions for $p_e^2c^2$ to obtain the result.

Solution 3.02

Problem 3.02

Have $\theta + \phi = 90$ degrees, so that:

$$\sin(\theta + \phi) = 1$$

$$\cos(\theta + \phi) = 0$$

Conservation of momentum in each direction:

$$E_i - E_f\cos\theta = cp_e\cos\phi$$

$$E_f\sin\theta = cp_e\sin\phi$$

Now apply the trigonometric identity:

$$\cos(\theta + \phi) = \cos\theta\cos\phi - \sin\theta\sin\phi$$

Multiply through the conservation of momentum equations by $\cos\theta$ and add them:

$$E_i\cos\theta - E_f\cos^2\theta = cp_e\cos\theta\cos\phi$$

$$E_f\sin^2\theta = cp_e\sin\theta\sin\phi$$

$$E_i\cos\theta - \left(E_f\cos^2\theta + E_f\sin^2\theta\right) = 0$$

$$E_i\cos\theta = E_f$$

Now put this back into the regular Compton equation to give:

$$\frac{m_ec^2}{E_f} = \frac{m_ec^2\cos\theta}{E_f} + 1 - \cos\theta$$

$$E_f = m_ec^2$$

Solution 3.03

Problem 3.03

The minimum photon energy will produce an electron and positron at rest, that is, $E_f = 2m_ec^2$. Use the Compton scattering formula derived in Problem 1 to get:

$$\frac{1}{h\nu_f} - \frac{1}{h\nu_i} = \frac{1}{m_ec^2}(1 - \cos\theta)$$

$$\frac{m_ec^2}{h\nu_f} - \frac{m_ec^2}{h\nu_i} = 1 - \cos\theta$$

$$\frac{1}{2} - \frac{m_ec^2}{E_i} = 1 - \cos\theta$$

$$\frac{m_ec^2}{E_i} = \cos\theta - \frac{1}{2}$$

The initial energy must of course be < 0, so we require $\cos\theta < \frac{1}{2}$.

Solution 3.04

An electron passing through an accelerating voltage V has a kinetic energy KE given by
$$KE = eV = (m - m_0)c^2 = (\gamma - 1)m_0c^2.$$

If phase velocity $\left(\frac{c^2}{v}\right)$ is twice the group velocity v, then:

$$v^2 = \frac{c^2}{2}$$

or

$$\gamma = \frac{1}{\sqrt{1 - \frac{1}{2}}} = \sqrt{2},$$

giving for voltage: $V = \left(\sqrt{2} - 1\right)\dfrac{mc^2}{e}.$

Solution 3.05

a. The easiest way to look at this is to use the Lorentz invariant, which says that $E^2 - c^2 p^2$ is the same in any frame. In the center of mass frame (CM), $p = 0$ initially, while in the lab frame, $p = p_0$ initially, where p_0 is in the incoming proton.

So, $E_{CM}^2 = E_{LAB}^2 - c^2 p_0^2 = \left((\gamma + 1)m_p c^2\right)^2 - c^2 p_0^2$, where m_p is the proton mass.

After the collision, the CM energy at threshold is simply $E_{CM}^2 = 4m_p c^2$.

This gives: $16m_p^2 c^4 = \left((\gamma + 1)m_p c^2\right)^2 - c^2 p_0^2$

Now use the identity (identical to $E = \gamma m c^2$):

$$\gamma^2 m_p^2 c^4 - c^2 p_0^2 = m_p^2 c^4$$

Plug into the above to give $\gamma = 7$.

So, the kinetic energy of the incoming proton is $6m_pc^2$.

b. Not possible as it violates baryon number conservation. This experiment helped confirm this conservation rule.

Solution 3.06 Problem 3.06

Conservation of momentum:

$$\vec{P}_\gamma = -\vec{P}_p$$

$$p_p^2 c^2 = E_p^2 - m_p^2 c^4 = E_\gamma^2$$

Conservation of energy:

$$mc^2 = E_p + E_\gamma = E_\gamma + \sqrt{E_\gamma^2 + m_p^2 c^4}$$

Plug in values, using $hc = 1239.8$ eVnm.

Solution 3.07 Problem 3.07

The red shift is given by:

$$\frac{\Delta \lambda}{\lambda} = \sqrt{\frac{1 + \frac{v}{c}}{1 - \frac{v}{c}}} - 1 \equiv z,$$ which can be arranged to give $\frac{v}{c} = \frac{(z+1)^2 - 1}{(z+1)^2 + 1}$.

The energy levels of hydrogen are given as a function of the Rydberg constant, R, as

$$\frac{1}{\lambda} = R\left(\frac{1}{n_1^2} - \frac{1}{n_2^2}\right) = R\left(\frac{1}{4} - \frac{1}{9}\right) = 0.139R$$

Since in our case, $\lambda = 654$ nm, we get $z = 0.153$ and $\frac{v}{c} = 0.14$.

Solution 3.08

Problem 3.08

The muons' effective lifetime will be dilated by $t' = \gamma t$ (or another way to think of it is that in the frame of the muon, the distance traveled is Lorentz contracted by the same amount). At the given velocity, $\gamma = \sqrt{1 - \frac{v^2}{c^2}} = 5$.

The amount of time it takes to travel 2000 m at 0.98 c is 6.8 μs, Lorentz contracted to $\frac{6.8}{5} =$ 1.4 μs or just about 1 half life. So about half should make it through.

Solution 3.09

Problem 3.09

If the proton has mass m, the electromagnetic force should fall off with distance r as $e^{-r\mu}$, where μ is the reduced photon mass with units m^{-1}.

$$\mu = \frac{mc}{\hbar}$$

$$\mu \ll \frac{1}{r}$$

$$mc^2 \ll \frac{\hbar c}{r} = \frac{\left(6.58 \times 10^{-16}\text{eVs}\right)\left(3 \times 10^8 \frac{\text{m}}{\text{s}}\right)}{5 \times 10^8 \text{m}} = 4 \times 10^{-16}\text{eV}$$

Solution 3.10

Problem 3.10

The difference in time arrival after traveling a distance L is given by:

$$\Delta t = L\left(\frac{1}{v_{g1}} - \frac{1}{v_{g2}}\right) \approx \frac{L}{8\pi^2 c}\left(\lambda_2^2 - \lambda_1^2\right)\mu^2$$

or

$$mc^2 < hc\sqrt{\frac{8\pi^2 c \Delta t}{L\left(\lambda_2^2 - \lambda_1^2\right)}}$$

Plug in the values: blue photon, 400 nm; red photon, 800 nm; 10^{-3} light-years = 9.46×10^{18} m.

Solution 3.11

The recoil energy of the nucleus must be matched by the Doppler shift of the photon caused by the motion of the nucleus towards it in order to get resonant absorption.

Recoil energy: $\dfrac{E_\gamma^2}{2mc^2}$, Doppler shift of photon energy: $E_\gamma \dfrac{v}{c}$, and matching the two gives:

$$v = \frac{E_0}{2mc}.$$

The photon would be a gamma ray. This has relevance to Mossbauer spectroscopy.

Solution 3.12

Problem 3.12

a. $\gamma t = 2.3$ weeks

b. $\gamma v T = 3.75 \times 10^{14}$ m

c. $0.5c$

Solution 3.13

Problem 3.13

a. The relationship between lifetime τ and linewidth T is simply $T\tau = h$, giving 1.43×10^{10} s.

b. Doppler shift is non-relativistic here, so:

$$\Delta E = \frac{v}{c} E_0$$

$$v = \frac{(4.6 \times 10^{-6}\text{eV})}{129 \times 10^3\text{eV}} c = 0.0107\frac{\text{m}}{\text{s}}$$

Solution 3.14

Look at the Lorentz invariant $E^2 - c^2p^2$. Initially $p_{CM} = 0$, so $E^2_{CM} = E^2_{LAB} - c^2p^2_{LAB}$.

After the collision, at threshold in the CM frame, momentum is 0 and energy is

$E_{CM} = m_\pi c^2 + m_e c^2$, giving:

$$(m_n + m_\pi)^2 c^4 = E^2_{LAB} - c^2p^2_{LAB} = \left(kT + E_p\right)^2 - c^2\left(\frac{kT}{c} + p_p\right)^2 = 2E_p kT - 2kTcp_p + m_p^2 c^4$$

Now the slightly tricky part is to solve this equation for p_p. For ease of manipulation, define $(m_n + m_\pi)^2 c^4 \equiv X$. Now have:

$$X \equiv 2kT\left(E_p - cp_p\right)$$

$$\left(X + 2kTcp_p\right)^2 = 4(kT)^2\left(m_p^2 c^4 + c^2 p_p^2\right)$$

$$X^2 + 4kTXcp_p = 4(kT)^2 m_p^2 c^4$$

$$4kTXcp_p = 4(kT)^2 m_p^2 c^4 - X^2$$

$$cp_p = \frac{kTm_p^2 c^4}{X} - \frac{X}{4kT}$$

So the final expression for the threshold is:

$$p_p \geq \frac{\left[(m_n + m_\pi)^2 - m_p^2\right]c^3}{4kT} - \frac{m_p^2}{(m_n + m_\pi)^2 - m_p^2}\frac{kT}{c}$$

Solution 3.15

Problem 3.15

This correction is called "fine structure." The energy difference ΔE is found by expanding the kinetic energy to first order in $\frac{p}{mc}$, which can be assumed to be small because the characteristic electron velocity is $\frac{c}{137}$. Doing this expansion gives:

$$T = \sqrt{p^2c^2 + m^2c^4} - mc^2$$

$$\rightarrow \frac{p^2}{2m}\left[1 - \frac{1}{4}\left(\frac{p}{mc}\right)^2 + \cdots\right]$$

so the correction to the Hamiltonian is $\Delta H = -\frac{p^4}{8m_0^3c^2}$ and the corrections to the energies for quantum numbers (l, m, n) are given by:

$$\Delta H_{nlm} = -\frac{1}{8m_0^3c^2}\langle nlm|p^4|nlm\rangle$$

We will not show all of the details of the calculation here (they are in the references), but it is readily shown that $\Delta E_0 = -\frac{E_0^2}{2m_0c^2} = -1.8 \times 10^{-4}$ eV.

Solution 3.16

Problem 3.16

The key here is to recognize that Lorentz contraction occurs only along the direction of motion. The projections along the x' and y' axes are $L\cos\theta$ and $L\sin\theta$, respectively. In the stationary observer's frame, the x projection is contracted as $L\cos\theta \rightarrow \frac{L\cos\theta}{\gamma}$.

The total length of the rod appears to be $L_{\text{obs}} = \sqrt{L^2\sin^2\theta + \frac{L^2\cos^2\theta}{\gamma^2}} = L\sqrt{1 - \frac{v^2}{c^2}\cos^2\theta}$.

In the observer's frame, the angle made with the x axis is given by $\tan\theta_{obs} = \gamma\tan\theta$, so the rod is rotated as well as contracted.

Solution 3.17

a. $L = K - V$

$$L = \frac{1}{2}m\left(\left(r\dot{\phi}\right)^2 + \dot{r}^2\right) - V(r)$$

$$\frac{d}{dt}\left(\frac{\partial L}{\partial \dot{r}}\right) = \frac{\partial L}{\partial r}$$

$$m\ddot{r} = mr\dot{\phi}^2 - \frac{\partial V(r)}{\partial r}$$

$$\frac{d}{dt}\left(\frac{\partial L}{\partial \dot{\phi}}\right) = \frac{\partial L}{\partial \phi}$$

$$\frac{d}{dt}\left(mr^2\dot{\phi}\right) = 0$$

This latter equation represents conservation of angular momentum.

b. $$-\int_0^{2\pi} G\frac{M_{ring}}{2\pi R}d\theta = -\int_0^{2\pi} G\frac{M_{ring}}{2\pi\sqrt{\left(r - h\cos\theta\right)^2 + \left(h\sin\theta\right)^2}}d\theta$$

$$= -\int_0^{\pi} G\frac{M_{ring}}{2\pi\sqrt{r^2 + h^2 - 2rh\cos\theta}}d\theta = -2\int_0^{\pi}\frac{GM_{ring}}{2\pi\sqrt{r^2 + h^2 - 2rh\cos\theta}}d\theta$$

c. Let $x = \frac{h}{r}$. Now

$$-2\int_0^{\pi}\frac{M_{ring}}{2\pi\sqrt{r^2 + h^2 - 2rh\cos\theta}}d\theta = \frac{-GM_{ring}}{\pi r}\int_0^{\pi}\frac{1}{\sqrt{1 + x^2 - 2x\cos\theta}}d\theta$$

Taylor expand:

$$\frac{\partial}{\partial x}\left[\frac{1}{\sqrt{1+x^2-2x\cos\theta}}\right] = \left(\frac{-1}{2}\right)\left(\frac{2x-2\cos\theta}{\left(1+x^2-2x\cos\theta\right)^{\frac{3}{2}}}\right)$$

$$\frac{\partial^2}{\partial x^2}\left[\frac{1}{\sqrt{1+x^2-2x\cos\theta}}\right] = \left(\frac{-1}{2}\right)\left(\frac{2}{1+x^2-2x\cos\theta^{\frac{3}{2}}}\right) + \left(\frac{-1}{2}\right)\left(\frac{-3}{2}\right)\left(\frac{(2x-2\cos\theta)^2}{1+x^2-2x\cos\theta^{\frac{5}{2}}}\right)$$

The power series then becomes:

$$= \frac{-GM_{ring}}{\pi r}\int_0^\pi \frac{1}{\sqrt{1+x^2-2x\cos\theta}}d\theta$$

$$= \frac{GM_{ring}}{\pi r}\int_0^\pi \frac{d\theta}{\sqrt{1+x^2-2x\cos\theta}}$$

$$= \frac{GM_{ring}}{\pi r}x\int_0^\pi \left[1+\left(\frac{-1}{2}\right)(-2\cos\theta)x+\left(\frac{1}{2}\right)\left(\frac{-1}{2}\right)\left[2+\left(\frac{-3}{2}\right)(-2\cos\theta)^2\right]x^2+\ldots\right]d\theta$$

$$= \frac{GM_{ring}}{\pi r}x\left[\pi+\left(\frac{-\pi}{2}+\frac{3}{2}\frac{\pi}{2}\right)x^2+\ldots\right] = -\frac{GM_{ring}}{r}\left[1+\frac{1}{4}\left(\frac{h}{r}\right)^2+\ldots\right]$$

This clearly goes to the expected value of zero as r goes to infinity.

d. $U = -M_P G\left[\frac{M_{bulge}}{r}\left(1+\frac{1}{4}\frac{R_{sun}^2}{r^2}\right)+\frac{M_{sun}}{r}\right]$

e. $\vec{F} = -\nabla U = -\frac{MG}{r^2}\left(M_{bulge}+M_{sun}\right)\hat{r} - \frac{3}{4}\frac{MG}{r^4}M_{bulge}R_{sun}^2\hat{r}$

f. $m\ddot{r} = mr\dot{\phi}^2 - \frac{\partial V(r)}{\partial r}$

$mr^2\dot{\phi} = L$

$\frac{d}{dt} = \frac{\partial\phi}{\partial t}\frac{d}{d\phi} = \frac{L}{mr^2}\frac{d}{d\phi} = \frac{Lu^2}{m}\frac{d}{d\phi}$

$$\frac{d^2}{dt^2} = \frac{Lu^2}{m}\frac{d}{d\phi}\left(\frac{Lu^2}{m}\frac{d}{d\phi}\right) = 2\frac{L^2u^3}{m^2}\frac{du}{d\phi}\frac{d}{d\phi} + \frac{L^2u^4}{m^2}\frac{d^2}{d\phi^2}$$

$$m\frac{\partial^2}{\partial t^2}r = m\frac{\partial^2}{\partial t^2}\frac{1}{u} = m\left[2\frac{L^2u^3}{m^2}\frac{du}{d\phi}\frac{d}{d\phi} + \frac{L^2u^4}{m^2}\frac{d^2}{d\phi^2}\right]\frac{1}{u}$$

$$= m\left[2\frac{L^2u^3}{m^2}\frac{du}{d\phi}\left(\frac{-1}{u^2}\right)\frac{du}{d\phi} + \frac{L^2u^4}{m^2}\left(\frac{2}{u^3}\left(\frac{du}{d\phi}\right)^2 + \frac{-1}{u^2}\frac{d^2u}{d\phi^2}\right)\right] = -\frac{Lu^2}{m}\frac{d^2u}{d\phi^2}$$

and for the other portion:

$$mr\dot{\phi}^2 - \frac{\partial V(r)}{\partial r} = \frac{m}{u}\left(\frac{Lu^2}{m}\right)^2 + F(r)$$

For: $-\dfrac{Lu^2}{m}\dfrac{d^2u}{d\phi^2} = \dfrac{m}{u}\left(\dfrac{Lu^2}{m}\right)^2 + F(r)$

$$\frac{d^2u}{d\phi^2} = -u - \frac{m}{u^2L^2}F(r)$$

$$F(u) = -MGu^2\left(M_{\text{bulge}} + M_{\text{sun}}\right) - \frac{3}{4}MM_{\text{bulge}}Gu^4R_{\text{sun}}{}^2$$

as expected.

g. First, try solving:

$$\frac{d^2u}{d\phi^2} + u = \frac{1}{\alpha}$$

Both homogeneous and particular solutions are necessary. Clearly, homogeneous solutions are of the form $u = c_1\cos\theta + c_2\sin\theta$. A particular solution is of the form $\frac{1}{\alpha}$, so the general solution is the sum $u = \frac{1}{\alpha} + c_1\cos\theta + c_2\sin\theta$.

One way to approximate the solution to the general equation would be to use the power-series method, writing the solution as $u = \sum_{x=0}^{\infty} c_x\theta^x$. This is complicated a bit by the presence of the u^2 term, so that in developing the recursion one uses $u^2 = \sum_{x=0}^{\infty}\sum_i c_x c_{x-i}\theta^x\left[i = \frac{x}{2}\to 1 \text{ else } 2\right]$. This gives the short-term solution.

Through analytic continuation, taking steps within the radius of convergence of this solution, one could extrapolate through one cycle in order to determine the offset of the next perihelion one cycle later. This gives the precession of the perihelion due to the bulge.

References
Special relativity

Original references to some of the problems in this chapter:

1. Compton, A. A Quantum Theory of the Scattering of X-Rays by Light Elements. *Physical Review* **212**, 483-502 (1923).

2. Davis, L., Goldhaber, A.S. & Nieto, M.M. Limit on the Photon Mass Deduced from Pioneer-10 Observations of Jupiter's Magnetic Field. *Phys. Rev. Lett.* **35** 1402-1405 (1975).

3. Dirac, P.A.M. The quantum theory of the electron. *Proc. R. Soc. London A* **117** 610-612 (1928).

4. Dirac, P.A.M. The quantum theory of the electron Part II. *Proc. R. Soc. London A* **118** 351-361 (1928).

5. Mossbauer, R. Recoilless Nuclear Resonance Absorption. *Annual Review of Nuclear Science* **12**, 123-152 (1962).

Also see:

6. Luo, J., Tu, L.C., Hu, Z.K. & Luan, E.J. New experimental limit on the photon rest mass with a rotating torsion balance. *Phys Rev Lett* **90**, 081801 (2003).

7. Tu, L., Luo, J. & Gillies, G.T. The mass of the photon. *Rep. Prog. Phys.* **68**, 77-130 (2005).

8. Lamb, W.E. & Retherford, R.C. Fine structure of the hydrogen atom by a microwave method. *Physical Review* **72**, 241-243 (1947).

9. Griffiths, D.J. *Introduction to Quantum Mechanics*, (Prentice Hall, Saddle River, NJ, 2004).

10. French, A.P. *Special Relativity*, (Massachussetts Institute of Technology, Cambridge, MA, 1968).

11. Resnick, R. *Introduction to Special Relativity* (John Wiley & Sons, Hoboken, NJ, 1968).

4

Quantum mechanics

Problems

Problems

Problems

Problem 4.01

Solution 4.01

The electronic binding energy in a nucleus is affected by the nuclear size and is most important for K electrons.

a. Use first-order perturbation theory to calculate a general formula for the change in energy ΔE of a K X-ray, assuming that screening effects can be neglected, the effect on the L electron is small, and the nuclear change distribution is uniform up to $R = 1.2A^{\frac{1}{3}}$.

b. Apply the formula to give a numerical value for lead (^{209}Pb, Z = 82).

c. Get a numerical value for hydrogen (Z = 1) due to the finite size of the proton.

Hint: The non-relativistic wave function for a point nucleus is:

$$U_0(r) = \frac{\gamma^{\frac{3}{2}}}{\sqrt{\pi}} \exp(-\gamma r)$$

$$\gamma = Z \frac{me^2}{\hbar^2}$$

The unperturbed energy is:

$$E_0 = -\frac{Z^2 me^4}{2\hbar^2} = -13.6Z^2 (eV)$$

Problem 4.02

Solution 4.02

The Geiger-Nuttal relation is an empirical law relating the decay rates of alpha-particle-emitting nuclei to the energies of the alpha-articles. In the '20s, George Gamow recognized that this relation resulted from the fact that alpha-decay is a quantum mechanical barrier-penetration problem. Work out the theory of this relation, showing that there is a linear relationship between $\ln\lambda$ (the decay constant related to the half-life) and $E^{\frac{1}{2}}$, where E is the total kinetic energy of the alpha-particle and the daughter nucleus.

Problem 4.03   Solution 4.03

An atom of an alkali metal has a single unpaired valence electron with quantum numbers n, l, j, j_z. It is now placed in a constant magnetic field pointing in the z direction. What are the quantum numbers describing its state if:

a. The magnetic energy is much smaller than the spin-orbit interaction?

b. The magnetic energy is large compared to the spin-orbit interaction?

c. Could the energy shift in either case be given by $\frac{e\hbar}{2mc}Bj_z$?

Problem 4.04   Solution 4.04

A system of two non-interacting spin-½ particles is known to have spin projections of $s_{1z} = s_{2z} = ½$. What is the probability that the total spin of the system is 1?

Problem 4.05   Solution 4.05

According to the Kronig-Penney model, an infinite one-dimensional crystal can be described as a periodic spacing of delta-function potentials with strength W and spacing a:

$$V(x) = \frac{-\hbar^2}{2m}W \sum_{n=-\infty}^{\infty} \delta(x - na)$$

(see the figure on the following page).

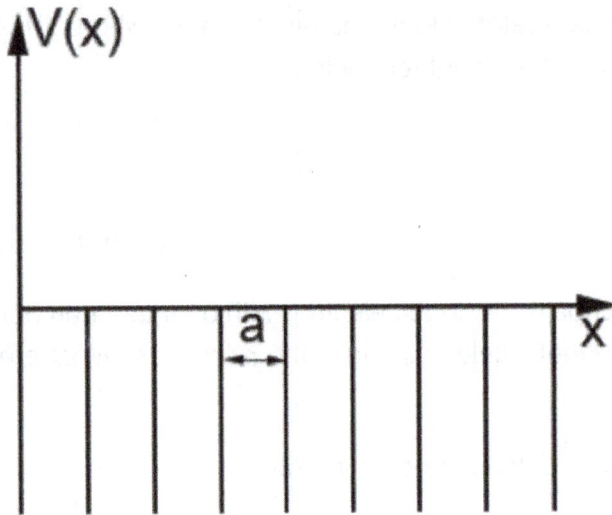

a. Find an expression for the general form of the eigenstates ψ of the one-electron Hamiltonian.

b. Between two delta functions, solutions are $Ae^{i\mu x} + Be^{-i\mu x}$, where $\mu = \dfrac{\sqrt{2mE}}{\hbar}$. Using boundary conditions, write down conditions on A, B, and μ.

c. Solve for B in terms of A to give the quantization conditions.

d. Discuss how this relates to the band theory of solids.

Problem 4.06

Solution 4.06

An electron moves in a one-dimensional harmonic oscillator potential of the form $\frac{1}{2}kx^2$. The system is placed on a constance E field pointing along the x direction.

a. What is the Hamiltonian?

b. Use perturbation theory to find an expression for the energy shift of the nth order eigenstate to first order in E.

c. Calculate the second-order shift (quadratic in E).

d. Show that it is possible to calculate the shift exactly and that it is equivalent to second-order perturbation theory.

Problem 4.07   Solution 4.07

Consider a charged particle scattered by a circular solenoid of radius a with a magnetic field B inside (Aharonov-Bohm effect). Formulate an expression for:

a. The phase difference between the particles at point B.

b. The form of the wave functions.

Hint: The wave function splits into right and left paths around the solenoid. All the right paths are identical, and all the left paths are identical.

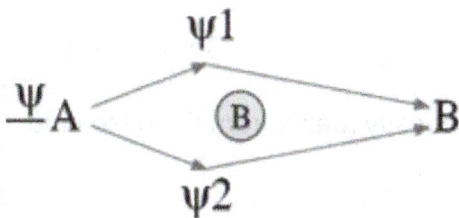

Problem 4.08   Solution 4.08

An ensemble of neutrons consists of a fraction f polarized in the $+z$ direction and a fraction $(1-f)$ polarized in the x direction.

a. Give the density matrix describing this system.

b. Using this result, find the ensemble probability of finding a neutron with spin in the $+z$ direction.

c. Calculate the ensemble average of s_z.

Problem 4.09   Solution 4.09

For a spin-1 particle, evaluate the matrix elements of $S_z(S_z + \hbar)(S_z - \hbar)$.

Problem 4.10   Solution 4.10

For a pure ensemble of spin-½ particles, is knowing the ensemble averages $\langle S_x \rangle$ and $\langle S_z \rangle$ sufficient to determine the state vector? If not, what additional information is needed?

Problem 4.11   Solution 4.11

For a mixed ensemble of spin-½ particles, construct the density matrix from the ensemble averages $\langle S_x \rangle$, $\langle S_y \rangle$, and $\langle S_z \rangle$.

Problem 4.12   Solution 4.12

For a mixed ensemble of spin-1 particles, how many parameters must be known?

Problem 4.13

Solution 4.13

In Schwinger's scheme for angular momentum, what are the "total raising operator" and "total lowering operator?" What are the non-zero matrix elements of each?

Problem 4.14

Solution 4.14

a. What is the 3×3 matrix J_y for a $j = 1$ system?

b. Write an expression for $\left\langle j = 1, m \left| J_y \right| j = 1, m' \right\rangle$.

Problem 4.15

Solution 4.15

The deuteron can be represented as a proton and a neutron in a 3-dimensional square well potential of depth V_0 and radius a, as shown in the figure on the following page.

a. Find the values of $V_0 a^2$ that just make the bound states with $\ell > 0$ possible.

b. Evaluate this in terms of MeV for reasonable deuteron values ($a = 2.00$ fm, $E_B(\ell = 0) = -2.23$ MeV, $m_p = 938.3$ MeV, $m_n = 939.6$ MeV) and compare the result with the "exact" value (~35 MeV).

c. Comment on the possibility of other nucleon-nucleon bound states.

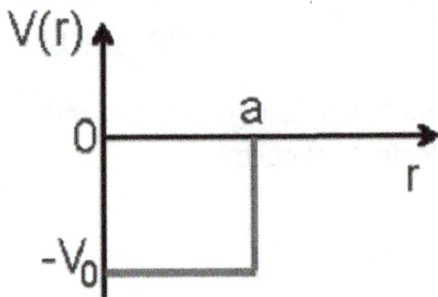

Problem 4.16

Solution 4.16

Find the energy levels and eigenfunctions for a particle in a two-dimensional circular potential, as shown. Under what conditions can a bound state exist? What is the importance of this result?

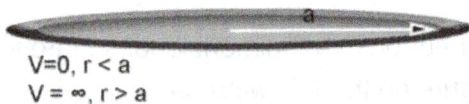

$V=0, r < a$
$V = \infty, r > a$

Problem 4.17

Solution 4.17

How can the eigenfunctions and eigenvalues of a 3D harmonic oscillator potential be obtained in terms of the solutions for the 1D oscillator?

Problem 4.18

Solution 4.18

A particle of mass m is contained in an infinite square-well potential of size $2L$ $(x = -L$ to $+L)$.

a. Calculate the ground state energy E_0.

b. If the walls are instantly moved out to $-2L$ and $+2L$ while the particle is in E_0, calculate the probability that it will remain the in ground state.

c. Calculate the probability that the particle will be in the first excited state after the expansion.

d. What is the expectation value of the energy?

e. If the box walls are expanded slowly, what is the probability that the particle will be in the ground state after expansion? What is the expectation value of the energy?

Problem 4.19
Solution 4.19

A particle of mass m is confined in an infinite square well potential with walls at $x = 0$ and $x = L$.

a. What are the energy eigenvalues and properly-normalized eigenfunctions?

b. Determine $\Delta x \Delta p$ for the nth eigenstate. Is this consistent with the Heisenberg uncertainty principle?

Problem 4.20
Solution 4.20

a. A particle is confined in an asymmetric box as shown in the figure. If there is an eigenstate at $E = E_0$, find an expression for the solutions for the width a.

b. Obtain a numerical value in the case where the particle is an electron $V_0 = 3$ eV and $E = 4$ eV.

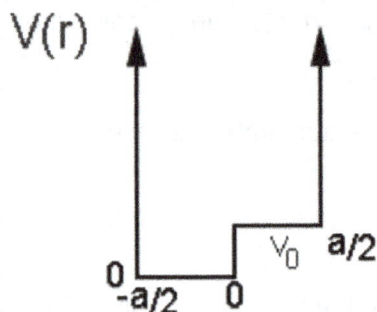

V(r)

0
-a/2 0 V_0 a/2

Problem 4.21

Solution 4.21

Two non-interacting spin-½ particles are placed in a 1D square well potential from $x = -L$ to $+L$.

a. What are the spin-space wave functions for the ground and first excited states?

b. If the particles interact via a repulsive Coulomb interaction, which of the degenerate first-excited states becomes lower in energy?

Problem 4.22

Solution 4.22

The nuclear electric quadrupole moment Q is given by

$$Q = e \left\langle j, m = j \left| \sum_{i+1}^{Z} \left(3z_i^2 - r_i^2 \right) \right| j, m = j \right\rangle.$$

a. Show that nuclei with $j = ½$ have $Q = 0$.

b. What does this mean for NMR studies?

c. Discuss how 9-fold degenerate n = 3 states of the hydrogen atom are split in an E field due to this effect. Treat the quadrupole moment as a perturbation and evaluate it to first order.

Problem 4.23

Solution 4.23

Born and Oppenheimer approximated the electronic, rotational, and vibrational energy levels as different order of the ratio $\left(\frac{m}{M}\right)^{\frac{1}{4}}$, where m is the electronic mass and M is the nuclear mass. Determine the order for each of these transactions; approximately what order of magnitude in energy does this correspond to?

Problem 4.24

Solution 4.24

What is the Heitler-London form of the internuclear potential for the hydrogen molecule (H_2)? What does it reduce to as $R \to$ infinity?

Problem 4.25

Solution 4.25

Derive the selection rules for rotational transitions of a diatomic molecule.

Problem 4.26

Solution 4.26

Discuss the selection rules for vibrational transitions of a diatomic molecule (homonuclear and heteronuclear).

Problem 4.27

Solution 4.27

The interaction between two spinless, uncharged atoms of mass m can be described using a Lennard-Jones potential.

a. What is the equation of this potential? Draw a sketch of its form.

b. Assuming small amplitude motion, derive an expression for the vibrational energy levels.

c. Derive an expression for the rotational energy levels of rotations around an axis perpendicular to a line connecting the nuclei. Assume the molecule is rigid (no stretching).

d. Where in the spectrum would you expect these energies to be?

Problem 4.28

Solution 4.28

Some of the rotational absorption lines of HCl occur at wavenumbers 83.03, 103.73, 124.30, 145.03, 165.51, and 185.86 cm^{-1}. What are the values of J to which these correspond? Estimate the separation distance between the H and Cl nuclei.

Problem 4.29

Solution 4.29

The energy required to remove an electron from a K atom is 4.32 eV. The energy required to remove the "extra" electron from Cl$^-$ is 3.80 eV.

a. At what nuclear distance does KCl become unbound?

b. If the equilibrium distance between the nuclei is 2.65 Å, estimate the number of rotational states in this spectrum.

Problem 4.30

Solution 4.30

The Ramsauer-Townsend effect predicts, purely quantum mechanically, that the scattering cross section for particles (such as electrons) off a 3-dimensional potential well (such as atoms of a noble gas) will vanish at some threshold energy.

a. Derive the expression for the differential and total scattering cross sections at small energy.

b. What is the value of $V_0 a^2$ for which the cross section vanishes at E = 0?

Problem 4.31   Solution 4.31

Derive a general expression for the phase shift created by a scattering potential $V(r) = \frac{A}{r^2}$, where A may be positive or negative.

Problem 4.32   Solution 4.32

Assume you have a Hermitian operator A for which $A^2 = A$.

a. What are the possible eigenvalues?

b. Show that the $\frac{1}{2}(1 \pm P)$ satisfies this equation, where P is the parity operator.

c. What happens when the operator in part (b) is applied to a wave function?

Problem 4.33   Solution 4.33

The lowest energy levels of carbon ($Z = 6$) are shown in the figure.

a. Describe what these energy levels are.

b. Derive the Landé g factor, and indicate how the energy levels would shift in the presence of a magnetic field.

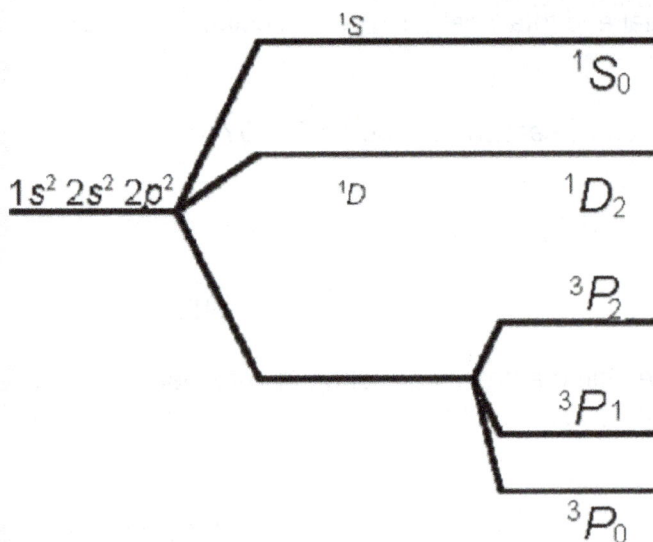

Problem 4.34

Solution 4.34

A particle with coordinates x, y, z has a wave function with the dependence $(x^2 - y^2)f(x^2 + y^2 + z^2)$. What L_z values are possible, and with what probability?

Problem 4.35

Solution 4.35

The spins of the two electrons in He can be in either a singlet state or a triplet state.

a. Explain which of the spin states will be found in the ground state and first excited state of this atom.

b. What would happen if the electrons were spinless bosons?

Problem 4.36

Solution 4.36

Low-density hydrogen may be excited to states of high n with large excitation energies ($\gg$ kT). Calculate the power radiated by transitions $n \rightarrow (n-1)$. Hint: Use the Larmor formula given by $\frac{dw}{dt} = \frac{2}{3}\frac{e^2 a^2}{c^3}$.

Problem 4.37

Solution 4.37

Estimate the force per unit area needed to pull apart a piece of perfect metal (cubic lattice).

Problem 4.38

Solution 4.38

The photoelectric emission from an alkali metal could be prevented by a potential of 1.07 V when the metal was irradiated with yellow (545 nm) light. When the wavelength was changed to violet (405 nm), the potential had to be increased to by an amount ΔV to prevent emission.

a. What is ΔV?

b. Estimate the work function of the material.

Problem 4.39

Solution 4.39

For a spin-½ system, what are the eigenvalues and (normalized) eigenvectors of $AS_y + BS_z$, where A and B are real constants?

Problem 4.40

Solution 4.40

A charge Z nucleus with a single orbiting electron undergoes beta decay: $n \rightarrow p + e^- + \bar{v}_e$. If the orbiting electron is initially in its ground state, what is the probability that it will still be in the ground state after beta decay?

Problem 4.41

Solution 4.41

The Yukawa potential can be used to represent the long-range potential of interaction between nucleons separated by a distance r,

$$V(r) = -\frac{g^2}{r} e^{\frac{-r}{R}},$$

where R and g are constants. Calculate the differential scattering cross section of a proton and a neutron in the Born approximation (assume the two nucleons have the same mass).

Problem 4.42

Solution 4.42

A stream of particles of mass m is moving in the x direction when it encounters a potential barrier of V_0 significantly higher than their kinetic energy E.

a. If the barrier has width a, find the transmission coefficient, that is, the probability of finding each particle at $x > a$ at a later time. What happens to this coefficient as Planck's constant approaches 0?

b. Calculate numerical values for electrons and protons for $E = 1$ eV and $V_0 = 2$ eV, $a = 1$ Å. Does this make sense?

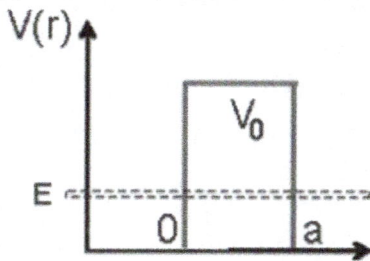

Problem 4.43

Solution 4.43

Classically, a particle sitting on a flat surface without moving has zero energy. Quantum mechanically, there is some fluctuation of its position above the surface. Calculate the minimum energy of a particle of mass m.

Problem 4.44

Solution 4.44

An electron is sitting on an infinite conducting surface. What is the Schrodinger equation that describes its motion, and what is the energy required to strip the electron off the surface?

Problem 4.45

Solution 4.45

Consider an array of N identical spinless bosons arranged as in the figure. The ring rotates about an axis perpendicular to the plane of the circle.

a. Write the Hamiltonian and solve for the eigenfunctions and eigenvalues.

b. What happens in the limit of a ring (i.e. $N \rightarrow$ infinity, but the mass remains finite)?

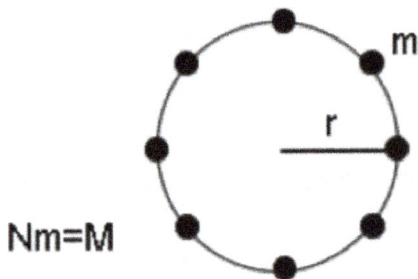

Nm=M

Problem 4.46

Assume that the moment of inertia of a nucleus, $\mathfrak{I}(I)$, changes slowly with I. Then for a given rotational band, show that:

a. $\dfrac{2\mathfrak{I}}{\hbar^2} \approx \dfrac{4I - 2}{E(I) - E(I - 2)}$

b. Show that the angular velocity can be expressed as $(\hbar\omega)^2 \approx 4I(I + 1)\left(\dfrac{E(I) - E(I - 2)}{4I - 2}\right)^2$

c. A plot of $\dfrac{2\mathfrak{I}}{\hbar^2}$ vs. $(\hbar\omega)^2$ for ^{160}Dy appears in the figure. Comment on the shape and what it means.

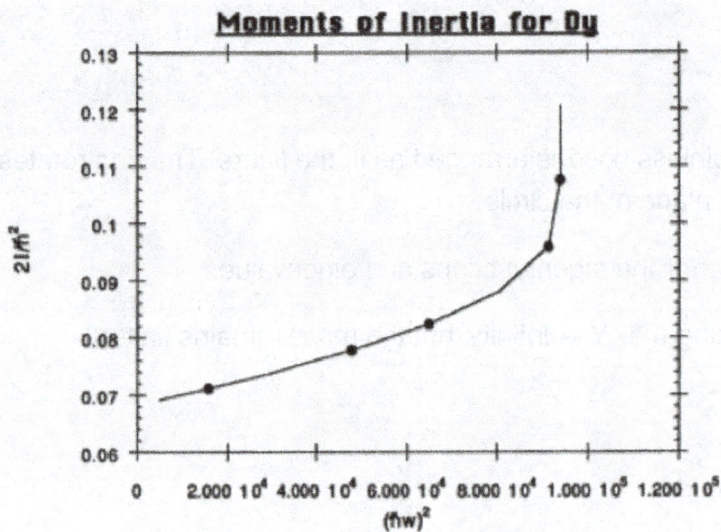

Moments of Inertia for Dy

Problem 4.47

In Problem 3.15, we saw that relativistic effects on the hydrogen atom lead to a "fine structure" correction to the Hamiltonian. In more detail, the fine structure terms are:

$$H_f = a\vec{p}^4 + \frac{b}{R^3}\left(\vec{L} \cdot \vec{S}\right) + W_D c\delta(R), \text{ where } a, b \text{ and } c \text{ are constants.}$$

Additional correction terms in the Hamiltonian, resulting in "hyperfine structure," result from interactions of the electron with the magnetic moment of the nucleus:

$$H_{hf} = d\left(\vec{L} \cdot \vec{I}\right) + \frac{e}{R^3}\left[3\left(\vec{S} \cdot \hat{n}\right)\left(\vec{I} \cdot \hat{n}\right) - \left(\vec{S} \cdot \vec{I}\right)\right] + f\left(\vec{S} \cdot \vec{I}\right)\delta\left(\vec{R}\right),$$

where d, e, and f are constants and I is the nuclear spin, S the electron spin, and L the angular momentum.

a. For each of the six terms multiplied by constants (a-f), state whether the term causes a shift in the energy of the 1s level, an energy splitting of the 1s level, both or neither. Explain your answer.

b. Express H_{hf} as a matrix in the basis set of eigenvectors common to

$$\vec{S}^2, \vec{I}^2, \vec{F}^2, F_z: \left\{\left| s = \tfrac{1}{2}; I = \tfrac{1}{2}; F; m_F\right\rangle\right\}$$ where $F = S + I$ is the total angular momentum, and thereby

derive the hyperfine splitting in terms of the constants d, e and f and spatial integrals involving the 1s orbital wave function ψ.

Solutions

Quantum mechanics

Solution 4.01

Problem 4.01

a. First calculate the potential energy of an electron inside and outside the area of uniform charge distribution, using the formula for an electron in a spherically symmetric potential distribution $\rho(r)$:

$$V(r) = -\frac{4\pi e}{R}\int_0^R \rho(r)r^2 dr + 4\pi e \int_R^\infty \frac{1}{r}\rho(r)r^2 dr$$

This gives:

$$V(r) = -\frac{Ze^2}{r}(r \geq R)$$

$$V(r) = -\frac{Ze^2}{r}\left(\frac{r^2}{2R^2} - \frac{3}{2}\right)(r \leq R)$$

Now use this to apply first-order perturbation theory to the region where the potential is perturbed— that is, for $r > R$:

$$\Delta E = \langle U_0|\Delta V|U_0\rangle$$

$$\Delta E = 4\gamma^3 \frac{Ze^2}{R}\int_0^R \left(\frac{r^2}{2R^2} - \frac{3}{2} + \frac{r}{R}\right)\exp(-2\gamma r)r^2 dr$$

This can be approximated assuming that $\gamma R \ll 1$. Do a Taylor series expansion; be sure to keep the first three terms in the exponential in order to have all of the first-order contributions. This will give:

$$\Delta E \approx \frac{4}{5} Z^2 E_0 \left(\frac{R}{a_0} \right)^2$$

$$a_0 = \frac{\hbar^2}{me^2}$$

b. 8.9 eV

c. 4×10^{-9} eV

Solution 4.02

Problem 4.02

First, realize that λ is proportional to T, where T is the tunneling amplitude (related by $\lambda = \frac{T v_0}{R}$, where v_0 is the velocity in the potential well).

Use the WKB approximation to calculate the tunneling amplitude:

$$T \sim \exp \left[-\frac{\sqrt{2m}}{\hbar} \int_R^{R_a} \sqrt{V(r) - E_a} \right],$$ where R is the radius of the bound particle and R_a is the

radius at which it becomes free. The energy and potential are given by the usual Coulomb potential (recall that $Z = 2$ for the alpha particle):

$$V(r) = \frac{2Ze^2}{4\pi\varepsilon_0 r}$$

$$E_a = \frac{2Ze^2}{4\pi\varepsilon_0 R_0}$$

The integral can be done exactly:

$$\int_R^{R_a} \sqrt{\frac{1}{r} - \frac{1}{R_a}} \, dr = \cos^{-1}\sqrt{\frac{R}{R_a}} - \sqrt{\frac{R}{R_a}\left(1 - \frac{R}{R_a}\right)} \approx \frac{\pi}{2} - 2\sqrt{\frac{R}{R_a}} \text{ for } R \ll R_a$$

So now:

$$-\ln(T) \approx \frac{\sqrt{2m}}{\hbar}\sqrt{\frac{2Ze^2}{4\pi\varepsilon_0}}\left(\frac{\pi}{2} - 2\sqrt{\frac{R}{R_a}}\right),$$ which gives the desired linear relationship with $E_a^{1/2}$.

Solution 4.03

Problem 4.03

a. $\left(n, l, s, j, j_z\right)$

b. $\left(n, l, l_z, s_z\right)$

c. No. In the first case, both orbit and spin must be considered. In the second case, j_z is not a good quantum number.

Solution 4.04

Problem 4.04

In order to have $s = 1$, we must have $s_{1z} = s_{2z} = \frac{1}{2}$. So we need the probability of $s_{2z} = \frac{1}{2}$ given that $s_{1x} = \frac{1}{2}$:

$$P = \left|\left\langle \Psi_{1/2}^{1/2}\middle|R\middle|\Psi_{1/2}^{1/2}\right\rangle\right|^2,$$ where R is the rotation operator:

$$P = \left|\left\langle \Psi_{1/2}^{1/2}\middle|e^{\frac{i}{\hbar}s_x}\middle|\Psi_{1/2}^{1/2}\right\rangle\right|^2 = \left[\exp\left(\frac{i}{\hbar}\alpha s_x\right)\right]^2,$$ where α is the angle of rotation. For $\alpha = 180°$,

$P = \frac{1}{2}$.

Solution 4.05

Problem 4.05

a. The general form is Bloch waves, which are a periodic function $u_k(x)$ multiplied by a plane wave $\exp(ikx)$. This can be shown by looking at the translation operator T that translates states by a distance a:

$$T\middle|\psi\rangle = \lambda\middle|\psi\rangle$$

$$\psi(x + a) = \langle x\middle|T\middle|\psi\rangle = \lambda\langle x\middle|\psi\rangle$$

$\lambda\psi(x) = \exp(ika)\psi(x)$

Now let:

$\psi(x) = u_k(x)\exp(ikx)$

$\exp(ika)\psi(x) = \psi(x + a) = u_k(x + a)\exp(ik[x + a]) \Rightarrow u_k(x + a) = u_k(x)$

b. For the first boundary condition, set $u_k(0) = u_k(a)$ to give:

$A\big[\exp(i\mu a - ika) - 1\big] + B\big[\exp(-i\mu a - ika) - 1\big] = 0$

For the second boundary condition, make the first derivative continuous:

$$\frac{\hbar^2}{2m}\left[\frac{d\psi(x)}{dx}\bigg|_{x=0} - \frac{d\psi(x)}{dx}\bigg|_{x=a}\right] = \frac{\hbar^2}{2m}W\psi(0)$$

$$\frac{\hbar^2}{2m}i\mu\Big[A - B - A\exp\big[i\mu a\big]B\exp\big[-i\mu a\big]\Big] = \frac{\hbar^2}{2m}W(A + B)$$

c. Solve for B in terms of A with the first boundary condition equation, then substitute into the second equation. Use the trigonometric identities:

$$\cos x = \frac{e^{ix} + e^{-ix}}{2}$$

$$\sin x = \frac{e^{ix} - e^{-ix}}{2i}$$

To give:

$$\cos ka = \cos\mu a + \frac{2W}{\mu}\sin\mu a$$

d. Plotting $\cos ka$ vs. μA gives bands of allowed states (see graph on following page):

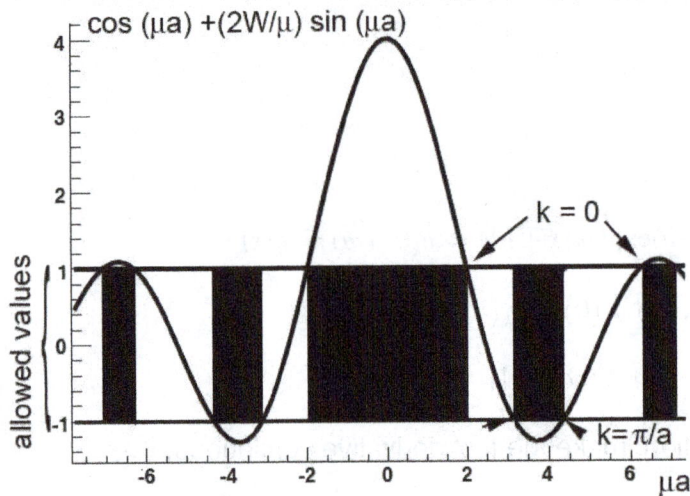

y-axis label: allowed values

Top curve label: cos (µa) +(2W/µ) sin (µa)

Labels: k = 0, k = π/a

x-axis: µa

Solution 4.06

Problem 4.06

a. $H = \dfrac{p^2}{2m} + \dfrac{1}{2}m\omega^2 x^2 - eEx$

b. The perturbation is $H' = -eEx$. Then by perturbation theory:

$$E_n = E_n^{(0)} + H'_{nm} + \sum_{m \neq n} \frac{\left| H'_{nm} \right|^2}{E_n^{(0)} - E_m^{(0)}} + \dots$$

To first order:

$$E_m^{(1)} + \left\langle n \left| H' \right| n \right\rangle = 0$$

c. To second order:

$$\sum_{m \neq n} \frac{\left| H'_{nm} \right|^2}{E_n^{(0)} - E_m^{(0)}} = \frac{\left| H'_{n+1,n} \right|^2}{E_n^{(0)} - E_{n+1}^{(0)}} + \frac{\left| H'_{n-1,n} \right|^2}{E_n^{(0)} - E_{n-1}^{(0)}}$$

Use $E_n^{(0)} - E_{n\pm 1}^{(0)} = \mp \hbar\omega$

Giving:

103

$$E_n^{(2)} - \frac{-eE^2}{\hbar\omega}\frac{\hbar}{2m\omega}(-n-1+n) = \frac{-e^2E^2}{2m\omega^2}$$

Solution 4.07 Problem 4.07

The Aharonov-Bohm effect is interesting because it is purely quantum mechanical— it is not predicted by classical electrodynamics because the key player is not the electric field but the electromagnetic vector potential. Inside the solenoid, the B field is nonzero; outside the B field is zero but the vector potential A is non-zero.

a. The right and left paths $S_{L,R}$ are identical:

$$\psi = \psi_1 + \psi_2 = \psi_1^0 \exp(-iS_R/\hbar) + \psi_2^0 \exp(-iS_L/\hbar)$$

And the difference between the paths gives the total magnetic flux:

$$\left. \int_B^A \vec{A} \cdot d\vec{r} \right|_R - \left. \int_B^A \vec{A} \cdot d\vec{r} \right|_L = \Phi$$

Using $\vec{F} = q\vec{v} \times \vec{B}$ simply gives $\delta = \frac{q}{\hbar}\Phi$ for the phase difference.

b. Now look at the Hamiltonian in terms of vector potential:

$$H = \frac{1}{2m}\left(\vec{p} - \frac{e}{c}\vec{A} \right)^2$$

In cylindrical coordinates:

$$\Delta^2 \psi = \frac{1}{\rho}\frac{\partial}{\partial \rho}\left(\psi \frac{\partial \psi}{\partial \rho} \right) + \frac{1}{\rho^2}\frac{\partial^2 \Omega}{\partial \varphi^2}$$

Now choose the gauge where:

$A_\rho = 0, A_\varphi = \frac{\Phi}{2\pi\rho}$, and rewrite the Hamiltonian as:

$$H = \frac{-\hbar^2}{2m} \left[\frac{\partial^2}{\partial\rho^2} + \frac{1}{\rho}\frac{\partial}{\partial\rho} + \frac{1}{\rho^2}\frac{\partial^2}{\partial\varphi^2} + \frac{2ie\Phi}{2\pi\hbar c}\frac{1}{\rho^2}\frac{\partial}{\partial\varphi} + \frac{e^2\Phi^2}{\hbar^2 c^2 4\pi^2\rho^2} \right]$$

Now define:

$$\alpha \equiv \frac{-e\Phi}{2\pi\hbar c}, \quad k \equiv \sqrt{\frac{2mE}{\hbar^2}},$$

and end up with the differential equation:

$$\left[\frac{\partial^2}{\partial\rho^2} + \frac{1}{\rho}\frac{\partial}{\partial\rho} + \frac{1}{\rho^2}\left(\frac{\partial}{\partial\varphi} + i\alpha\right)^2 + k^2 \right]\psi = 0$$

This equation may be separated into two parts: a part F dependent upon φ only; and a part R dependent upon ρ only.

$$\psi = F(\varphi)R(\rho)$$

$$F(\varphi)R'' + \frac{F}{\rho}R' + \frac{R}{\rho^2}\left(F'' + 2i\alpha F' - \alpha^2 F\right) + k^2 FR = 0$$

Solve for R:

$$aR'' + \frac{a}{\rho}R' + \left(\frac{b}{\rho^2} + k^2 a\right) = 0,$$

where a, b are functions of F only. The solutions to this are Bessel equations of order $\sqrt{-b/a}$.

Solve for F:

$$\left(D^2 + m^2\right)F = 0$$

$F = \exp(im\psi)$, where m is constant. So the solution is a sum of positive and negative order Bessel functions, but only the positive orders survive because they should have $\psi \to 0$ as $\rho \to 0$. So the solution may be expressed as:

$$\psi = \sum_{m=-\infty}^{\infty} \exp(im\varphi) A_m J_{|m+a|}(k\rho)$$

Note the asymptotic limit:

$$J_{\alpha+1}(r) \xrightarrow{r\to\infty} \sqrt{\frac{2}{\pi r}} \cos\left[r - \left(\frac{\alpha+1}{2} \right) \frac{\pi}{4} \right]$$

Solution 4.08

Problem 4.08

a. $\rho = f \left| s_z+ \right\rangle \left\langle s_z+ \right| + (1-f) \left| s_x+ \right\rangle \left\langle s_x+ \right|$

b. $P(z+) = \left\langle s_z+ \left| \rho \right| s_z+ \right\rangle = f + (1-f) \left| \left\langle s_z+ \left| s_x+ \right\rangle \right|^2$

$\quad = f + \dfrac{1}{2}\left(1-f\right) = \dfrac{1}{2}\left(1+f\right)$

c. $[s_z] = Tr(\rho s_z) = f\left(\dfrac{\hbar}{2}\right)$

Solution 4.09

Problem 4.09

$$S_z(S_z + \hbar)(S_z - \hbar)$$

We want to evaluate:

$$\left\langle j'm' \left| S_z(S_z + \hbar)(S_z - \hbar) \right| jm \right\rangle$$

Use:

$$S_z \left| jm \right\rangle = m\hbar \left| jm \right\rangle$$

$$\left\langle j'm' \left| m\hbar \right| jm \right\rangle = \delta_{jj'mm'}$$

$$\left\langle j'm' \left| S_z (S_z + \hbar)(S_z - \hbar) \right| jm \right\rangle = \left\langle j'm' \left| (S_z^2 + \hbar S_z)(S_z - \hbar) \right| jm \right\rangle$$

$$= \left\langle j'm' \left| S_z^3 - \hbar^2 S_z \right| jm \right\rangle = 0 \text{ for } m, j \neq m', j'$$

So we must have:

$$S_z^3 = \hbar^2 S_z$$

Solution 4.10

For a spin-½ particle, the state vector can be written in terms of the sigma matrices, which are a complete set of 2 × 2 matrices:

$$P_\alpha = a_0 + \vec{a} \cdot \vec{\sigma}$$

$Tr(P_\alpha) = 1 \Rightarrow a_0 = \frac{1}{2}$ because $Tr(\sigma) = 0$. Then use:

$$P_\alpha^2 = a_0^2 + 2a_0 \vec{a} \cdot \vec{\sigma} + \frac{\left(\vec{a} \cdot \vec{\sigma} \right)^2}{a^2} = P_\alpha$$

We know a_0, so this gives:

$$P_\alpha = \frac{1}{2}\left(1 + \hat{\pi}_\alpha \cdot \vec{\sigma} \right), \text{ where } \pi \text{ is the polarization vector, given by:}$$

$$\hat{\pi}_\alpha = \left\langle \alpha \left| \vec{\sigma} \right| \alpha \right\rangle = Tr(\sigma P_\alpha) = \frac{1}{2}Tr(\sigma) + \frac{1}{2}Tr\left[\left(\hat{\pi}_\alpha \cdot \vec{\sigma} \right) \vec{\sigma} \right]$$

The polarization vector is known if we know $\langle S_x \rangle$ and $\langle S_z \rangle$ and the sign of $\langle S_y \rangle$.

Solution 4.11

Spin may be up or down with probability p_α, p_β. Then:

$$\rho = \sum_\alpha \left| \alpha \right\rangle p_\alpha \left\langle \alpha \right| = p_\alpha P_\alpha + p_\beta P_\beta = p_\alpha \left(\frac{1}{2} + \hat{\pi}_\alpha \cdot \vec{\sigma} \right) + p_\beta \left(\frac{1}{2} - \hat{\pi}_\alpha \cdot \vec{\sigma} \right)$$

$$= \frac{1}{2} \left(1 + \left(p_\alpha - p_\beta \right) \hat{\pi}_\alpha \cdot \vec{\sigma} \right)$$

And so calculate:

$$\langle S \rangle = Tr\left(\rho \vec{\sigma} \right) = Tr\left[\frac{1}{2} \left(1 + \left(p_\alpha - p_\beta \right) \hat{\pi}_\alpha \cdot \vec{\sigma} \right) \cdot \vec{\sigma} \right] = \left(p_\alpha - p_\beta \right) \hat{\pi}_\alpha$$

So if all components of $\langle S \rangle$ are known, we can construct ρ.

Solution 4.12

Problem 4.12

A pure state of spin-1 can be represented by a column matrix with 3 rows, described by $4s = 4$ independent parameters. The projection operator can be written in terms of $(2s + 1)^2 = 9$ matrices (the lambda matrices, corresponding to spin, dipole, and quadrupole moments). For a mixed ensemble, it's necessary to know all the average components of spin, as well as those of the dipole and quadrupole moments, minus 1 due to normalization. This gives <u>eight</u> parameters.

Solution 4.13

Problem 4.13

Total raising operator:

$$K_+ = a_+^\dagger a_-^\dagger$$

Total lowering operator:

$$K_- = a_+ a_-$$

Matrix elements:

$$\langle n' | K_- | n \rangle = \langle n'_+ n'_- | a'_+ a'_- | n_+ n_- \rangle = \left\langle n'_+ n'_- \left| a_+^\dagger \sqrt{n_- + 1} \right| n'_+ n'_- + 1 \right\rangle$$

$$= \sqrt{(n_- + 1)(n_+ + 1)} \, \delta_{n_+, n_+ - 1} \delta_{n_-, n_- + 1}$$

Similarly:

$$n' | K_- | n = \sqrt{(n_-)(n_+)} \, \delta_{n_+, n_+ + 1} \delta_{n_-, n_- + 1}$$

Solution 4.14

Problem 4.14

a. $J_y = \begin{pmatrix} \langle + | J_y | + \rangle & \langle + | J_y | 0 \rangle & \langle + | J_y | - \rangle \\ \langle + | J_y | + \rangle & \langle 0 | J_y | 0 \rangle & \langle 0 | J_y | - \rangle \\ \langle - | J_y | + \rangle & \langle - | J_y | 0 \rangle & \langle - | J_y | - \rangle \end{pmatrix} = \dfrac{\hbar}{\sqrt{2}} \begin{pmatrix} 0 & -i & 0 \\ i & 0 & -i \\ 0 & i & 0 \end{pmatrix}$

b. $\dfrac{\hbar}{\sqrt{2}} \left(\delta_{m',1} \delta_{m',0} \delta_{m',-1} \right) \begin{pmatrix} 0 & -i & 0 \\ i & 0 & -i \\ 0 & i & 0 \end{pmatrix} \begin{pmatrix} \delta_{m,1} \\ \delta_{m,0} \\ \delta_{m,-1} \end{pmatrix}$

Solution 4.15

Problem 4.15

a. For a spherically symmetric potential:

$$-\frac{\hbar^2}{2m} \frac{d^2 u}{dr^2} + \left\{ V(r) + \frac{\ell(\ell + 1)\hbar^2}{2mr^2} \right\} u(r) = E u(r)$$

For zero angular momentum:

$$-\frac{\hbar^2}{2m} \frac{d^2 u}{dr^2} + a u(r) = E u(r) \text{ for } r < a$$

$$-\frac{\hbar^2}{2m}\frac{d^2u}{dr^2} = Eu(r) \text{ for } r > a$$

Define a bound state with $E = -E_B$, and solve the equations in both regions.

For $r < a$:

$$u(r) = A\sin(kr) + B\cos(kr)$$

$$k \equiv \frac{2m}{\hbar^2}(V_0 - E_B)$$

Impose the condition:

$$u(r) \xrightarrow{r \to 0} 0$$

$$\Rightarrow u(r) = A\sin(kr)$$

For $r > a$:

$$u(r) = C\exp(-k'r)$$

$$k \equiv \frac{2mE_B}{\hbar^2}$$

Set equations and derivatives equal at $r = a$:

$$A\sin(ka) = C\exp(-k'a)$$

$$kA\cos(ka) = -k'C\exp(-k'a)$$

Take the ratio of both sides:

$$\cot(ka) = -\sqrt{\frac{E_B}{V_0 - E_B}}$$

We just barely have a bound state when $E_B = 0$, or $\cot(ka) = 0$, or $ka = \frac{\pi}{2}$, or $V_0 a^2 = \frac{\hbar^2 \pi^2}{8m}$.

b. About 25 MeV, or a bit smaller than the "exact" value.

c. The neutron-neutron and proton-proton systems do not have a bound state because, being identical fermions, they must have antiparallel spins (singlet state). In this case, the depth of the potential well is insufficient to create a bound state. The bound state of the deuteron has $S = 1$. (Note that this has nothing to do with proton-proton electrostatic repulsion!)

Solution 4.16

Solve the 2-D Schrodinger equation with the boundary condition that the wave function vanishes at $r = a$.

$$\frac{-\hbar^2}{2m} \frac{1}{r} \frac{d}{dr}\left(r \frac{d\psi}{dr}\right) + V\psi = E\psi$$

$$\frac{d^2\psi}{dr^2} + \frac{1}{r}\frac{d\psi}{dr} + \frac{2mE}{\hbar^2}\psi = 0$$

Define:

$$\alpha = \frac{\sqrt{2mE}}{\hbar}$$

$$\rho = \alpha r$$

Then:

$$\alpha^2 \frac{d^2\psi}{d\rho^2} + \frac{\alpha^2}{\rho}\frac{d\psi}{d\rho} + \alpha^2\psi = 0$$

$$\frac{d^2\psi}{d\rho^2} + \frac{1}{\rho}\frac{d\psi}{d\rho} + \psi = 0$$

This is just Bessel's equation, so:

$$\psi = A J_m(\alpha r)$$

With the boundary condition:

$$J_m(\alpha a) = 0$$

So α are the 0s of the Bessel functions:

$$E_{n,m} = \frac{\hbar^2}{2ma^2} \left[0(J_m) \right]^2$$

Energies with $m \neq 0$ are degenerate. These solutions have been visualized in a famous scanning tunneling microscopy study that used iron atoms to "corral" surface state electrons.

Solution 4.17

Problem 4.17

The Hamiltonian is simply the sum of the components for each dimension. The eigenfunctions are the products of the eigenfunctions of each dimension:

$$\psi(xyz) = h_{n_x}\left(\sqrt{\frac{m\omega}{\hbar}}x \right) h_{n_y}\left(\sqrt{\frac{m\omega}{\hbar}}y \right) h_{n_z}\left(\sqrt{\frac{m\omega}{\hbar}}z \right) \exp\left(\frac{-m\omega}{2\hbar}[x^2 + y^2 + z^2] \right)$$

The eigenvalues are given by:

$$\hbar\omega\left(n + \frac{1}{2} \right) \text{ for each component, or, in total, } \hbar\omega\left(n_x + n_y + n_z + \frac{3}{2} \right) = \hbar\omega\left(N + \frac{3}{2} \right)$$

This is readily generalizable to N dimensions. Note that there is degeneracy in the $n > 0$ states in 3D.

Solution 4.18

Problem 4.18

a. Ordinary particle in a box energies with width $2L$:

$$E_0 = \frac{\hbar^2\pi^2}{8mL^2}$$

b. Probability of ground state after expansion:

$$\left| \langle \psi_0(4L) | \psi_0(2L) \rangle \right|^2 = \left| \frac{1}{\sqrt{L}} \frac{1}{\sqrt{2L}} \int_{-2L}^{2L} dx \cos\left(\frac{\pi x}{2L} \right) \cos\left(\frac{\pi x}{4L} \right) \right|^2 = \frac{4}{9\pi^2}$$

c. $P = 0$, since the integral vanishes.

d. E_0

e. $P = 1$, since the particle remains in the ground state. The energy is $E_0 = \dfrac{\hbar^2 \pi^2}{32mL^2}$.

Solution 4.19

Problem 4.19

a. $E_n = \dfrac{n^2 \pi^2 \hbar^2}{2mL^2}$

$$\psi_n(x) = \sqrt{\frac{2}{L}} \sin\left(\frac{n\pi x}{L}\right)$$

b. The uncertainties in the variables are given by the square roots of the variances, that is:

$$\Delta x = \sqrt{\langle x^2 \rangle - \langle x \rangle^2}$$

$$\Delta x = \sqrt{\langle p^2 \rangle - \langle p \rangle^2}$$

Knowing the wave functions, we can calculate these expectation values, knowing that the momentum operator is given by:

$$\hat{p} = \frac{\hbar}{i}\frac{\partial}{\partial x}$$

So this gives:

$$x = \frac{2}{L}\int_0^L x\sin^2\left(\frac{n\pi x}{L}\right) = \frac{L}{2}$$

$$x^2 = \frac{2}{L}\int_0^L x^2\sin^2\left(\frac{n\pi x}{L}\right) = \left(\frac{L}{2\pi n}\right)^2\left(\frac{4\pi^2 n^2}{3} - 2\right)$$

113

$$p = -i\hbar \frac{2\pi n}{L^2} \int_0^L \sin\left(\frac{n\pi x}{L}\right)\cos\left(\frac{n\pi x}{L}\right) = 0$$

$$p^2 = \frac{2\pi^2 n^2 \hbar^2}{L^3} \int_0^L \sin^2\left(\frac{n\pi x}{L}\right) = \frac{\pi^2 n^2 \hbar^2}{L^2}$$

So:

$$\Delta p = \frac{n\pi\hbar}{L}$$

$$\Delta x = \sqrt{\frac{L^2}{12}\left[1 - \frac{6}{n^2\pi^2}\right]}$$

Giving:

$$\Delta x \Delta p = \hbar\sqrt{\frac{n^2\pi^2 - 6}{12}}$$

This is consistent with the uncertainty principle, because the smallest this value can be is about 0.57, which is larger than the ½ required by the principle.

Solution 4.20

Problem 4.20

a. The solutions to this potential are found by writing the wave functions in each region and then matching them and their derivatives at the boundary conditions.

$$k \equiv \sqrt{\frac{2mE}{\hbar^2}}$$

$$q \equiv \sqrt{\frac{2m(E - V)_0}{\hbar^2}}$$

$\psi_{\mathrm{I}} = A\sin k\left[x + \frac{a}{2}\right]$ in left half of well

$\psi_{\mathrm{II}} = B\sin q\left[x - \frac{a}{2}\right]$ in right half of well

Match the functions and derivatives at $x = 0$ to give:

$$A\sin\frac{ka}{2} = -B\sin\frac{qa}{2}$$

$$kA\cos\frac{ka}{2} = Bq\cos\frac{qa}{2}$$

$$q\tan\frac{ka}{2} = -k\tan\frac{qa}{2}$$

b. This gives the form:

$$\tan\left(2\sqrt{\frac{2m}{\hbar^2}}\frac{a}{2}\right) = -2\tan\left(\sqrt{\frac{2m}{\hbar^2}}\frac{a}{2}\right)$$

Use the identity:

$$\tan 2x = \frac{2\tan x}{1 - \tan^2 x}$$

To give:

$$\frac{qa}{2} = n\pi \pm \arctan\sqrt{2}$$

$$a = 2.46\left[n\pi \pm \arctan\sqrt{2}\right] \text{ nanometers}$$

Solution 4.21

Problem 4.21

a. The wave functions are given by the following.

Ground state:

$$\cos\left(\frac{\pi x_1}{2L}\right)\cos\left(\frac{\pi x_2}{2L}\right)\chi_{m_s=0}^{s=0}(\sigma_1, \sigma_2)$$

First excited state:

$$\left[\cos\left(\frac{\pi x_1}{2L}\right)\sin\left(\frac{\pi x_2}{2L}\right) + \cos\left(\frac{\pi x_2}{2L}\right)\sin\left(\frac{\pi x_1}{2L}\right)\right]\left\{\begin{array}{c}\chi_{m_s=0}^{s=0}(\sigma_1,\sigma_2) \\ \chi_{m_s=0,\pm1}^{s=1}(\sigma_1,\sigma_2)\end{array}\right\}$$

b. For any repulsive interaction, the state in which the particles spend the most time farther apart will be lower in energy. In this case, this is the $s = 1$ state, which is spin-symmetric/spatially antisymmetric.

Solution 4.22

Problem 4.22

a. The quadrupole moment can be expressed in terms of spherical harmonics:

$$3z^2 - r^2 = \sqrt{\frac{16\pi}{5}}r^2 Y_2^0$$

Then the expectation value will be an integral over the $(\text{wave function})^2$ times this function, or the product of $Y_{2j}^m Y_2^0$. Since the spherical harmonics are mutually orthonormal, this vanishes for $j \leq \frac{1}{2}$.

b. Nuclei with $j = \frac{1}{2}$ are ideal for NMR studies, since they will show sharp peaks due to lack of a quadrupole moment.

c. The perturbation to the Hamiltonian will be proportional to Y_2^0. The $\ell = 1$ states do not couple to $\ell = 0$ or $\ell = 2$ because of parity. The $\ell = 1$ states are split into three states $m = -1, 0, 1$ according to:

$$\Delta_m(\ell = 1) \propto \left\langle n, 1, m \left| Y_2^0 \right| n, 1, m \right\rangle$$

Similarly:

$$\Delta_m(\ell = 2) \propto \left\langle n, 2, m \left| Y_2^0 \right| n, 2, m \right\rangle, \text{ except for the case } m = 0, \text{ which couples to } \ell = 0, \text{ and}$$

so much be considered separately, taking into consideration the coupling

$$\Delta = \left\langle n, 2, 0 \left| Y_2^0 \right| n, 0, 0 \right\rangle$$

Diagonalize:

$$H = \begin{pmatrix} \Delta_0(\ell = 2) & \Delta \\ \Delta & 0 \end{pmatrix}$$

$$\lambda^2 - \lambda\Delta_0 - \Delta^2 = 0$$

$$\lambda = \frac{1}{2}\Delta_0 \pm \sqrt{\frac{\Delta_0^2}{4} + \Delta^2}$$

Qualitatively:

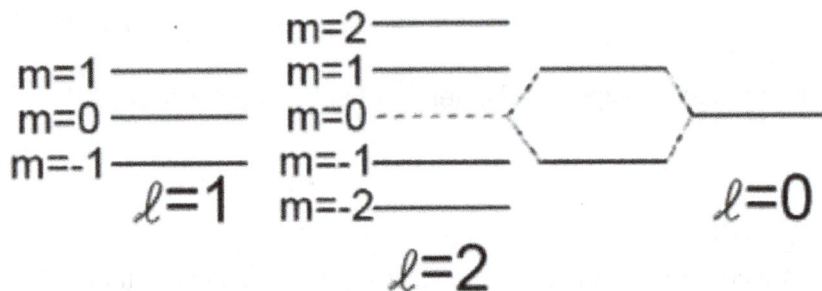

Solution 4.23

This is the basis of the Born-Oppenheimer approximation— not to be confused with the Born approximation and also called the adiabatic approximation. This model decouples the types of motion in a molecule under the assumption that the velocities of electrons are much greater than those of nuclei.

For electronic motion, the energy associated with a transition is going to be on the order of magnitude of the kinetic energy of an electron moving in a potential of radius equal to the dimensions of the molecule r, or:

$$E_{el} \sim \frac{\hbar^2}{mr^2}$$

For vibrational motion, the energies are those of a harmonic oscillator of spring constant k and mass M or:

$$E_{vib} \sim \hbar\sqrt{\frac{k}{M}}$$

Here's the tricky part: how to relate k to the dimensions of the molecule? A clever way to do it is to realize that a displacement equal to r should be on the order of the electronic energy, since it would deform the molecule. We can thus approximate:

$$k \sim \frac{E_{el}}{r^2}$$

$$E_{vib} \sim \sqrt{\frac{m}{M}} E_{el}$$

Finally, the rotational energies are simply related to the moment of inertia of the molecule, which is related to Mr^2:

$$E_{rot} \sim \frac{\hbar^2}{I} \sim \frac{m}{M} E_{el}$$

So electronic transitions are zeroth order in the parameter, vibrational transitions 2nd order, and rotational transitions 4th order. The value of $\frac{m}{M}$ is about 0.0001 - 0.001.

Solution 4.24

Problem 4.24

Heitler and London investigated the hydrogen molecule using quantum mechanics only a year after the Schrodinger equation was proposed! They postulated that the wave function of the molecule should be a linear superposition of the wave functions of electron a and electron b:

$$\Psi = c_1 \Psi_a(1) \Psi_b(2) + c_2 \Psi_a(2) \Psi_b(1)$$

Minimizing the energy leads to:

$$E \leq 2E_0 + \frac{\int \psi_a H \psi_a d\tau + \int \psi_a H \psi_b d\tau}{1 + \int \psi_1 \psi_2 d\tau}$$

The first term in the numerator is the Coulomb integral, and the second is the exchange integral. The term in the denominator is the overlap integral, measuring the amount that the two atoms overlap.

As $R \to \infty$: At large separations there is no chemical bond, so the overlap integral vanishes. This makes it possible to treat the system reasonably simply by treating the interaction between the two atoms as a perturbation. Assume both atoms are in the ground state:

$$H = H_{0a} + H_{0b} + H'$$

$$H = e_2 \left[\frac{1}{R} - \frac{1}{r_{2a}} - \frac{1}{r_{1b}} + \frac{1}{r_{12}} \right]$$

Take $R \to \infty$, so expand the other coordinates in powers of $\frac{1}{R}$:

$$\frac{1}{r_{1b}} = \frac{1}{\left| \vec{r}_1 - \vec{R} \right|} = \frac{1}{R} \frac{1}{\sqrt{1 - \left(\frac{2z_1}{R} \right) + \left(\frac{r^1}{R} \right)^2}} \approx \frac{1}{R} \left[1 + \frac{z_1}{R} - \frac{1}{2} \left(\frac{r_1}{R} \right)^2 + \frac{3}{2} \left(\frac{z_1}{R} \right)^2 \right]$$

$$\frac{1}{r_{2a}} = \frac{1}{\left| \vec{r}_2 + \vec{R} \right|} = \frac{1}{R} \frac{1}{\sqrt{1 + \left(\frac{2z_2}{R} \right) + \left(\frac{r_2}{R} \right)^2}} \approx \frac{1}{R} \left[1 - \frac{z_2}{R} - \frac{1}{2} \left(\frac{r_2}{R} \right)^2 + \frac{3}{2} \left(\frac{z_2}{R} \right)^2 \right]$$

$$\frac{1}{r_{12}} = \frac{1}{\left| \vec{r}_1 - \left(\vec{r}_2 + \vec{R} \right) \right|} = \frac{1}{R} \frac{1}{\sqrt{1 + \left(\frac{2(z_2 - z_1)}{R} \right) + \left(\frac{\vec{r}_2 - \vec{r}_1}{R} \right)^2}}$$

$$\approx \frac{1}{R} \left[1 - \frac{z_2 - z_1}{R} - \frac{1}{2} \left(\frac{\vec{r}_2 - \vec{r}_1}{R} \right)^2 + \frac{3}{2} \left(\frac{z_2 - z_1}{R} \right) \right]$$

Add them all up to give:

$$H' = \frac{e^2}{R^3} \left[x_1 x_2 + y_1 y_2 - 2 z_1 z_2 \right]$$

Note that this is just the dipole-dipole interaction, and could have been derived by expanding H' in a multipole expansion. So physically, the first order corresponds to the dipole-dipole interaction, the next order to dipole-quadrupole, and so on. Given this, it is easy to guess what the radial dependence will be, but let's keep going a little longer anyhow.

The first order perturbation vanishes because H' is an odd function of the coordinates (this is true for higher-order terms as well). So looking at the form of the second order perturbation:

$$\Delta E = \sum_{n\ell m} \frac{\left| \langle 100 | \langle 100 | H' | n_a \ell_a m_a \rangle | n_b \ell_b m_b \rangle \right|^2}{E_0 - E_a - E_b}$$

Since H' goes as R^{-3}, it follows that the first non-vanishing term goes as R^{-6}, which is of course the dipole-dipole term in the van der Waals equation. Now that we know this form, we can use it to determine semi-empirical potentials.

Solution 4.25

Problem 4.25

Allowed transitions are those whose matrix elements of the total dipole moment do not vanish. Since this operator is odd, the final states must have opposite parity. For rotational motion, the eigenvalue equation is given in terms of spherical harmonics:

$$\hat{H} Y_\ell^m(\theta, \phi) = \frac{\hbar^2}{2I} \ell(\ell + 1) Y_\ell^m(\theta, \phi), \text{ where } I \text{ is the moment of inertia of the molecule.}$$

Transition matrix elements are determined from the orthogonality of the spherical harmonics and thus will require $\ell' = \ell \pm 1$.

Solution 4.26

Problem 4.26

Vibrations can be approximated as a harmonic oscillator, so the selection rule should be the same as that of a harmonic oscillator, namely $\nu' = \nu \pm 1$.

But homonuclear molecules do not have a permanent dipole moment, so only heteronuclear ones will show a vibrational spectrum. Note that anharmonicity will change the selection rule and make certain other transitions weakly allowed.

Solution 4.27

a. $V(r) = V_0 \left[\left(\dfrac{r_0}{r} \right)^{12} - 2 \left(\dfrac{r_0}{r} \right)^6 \right]$, where $V_0, r_0 > 0$.

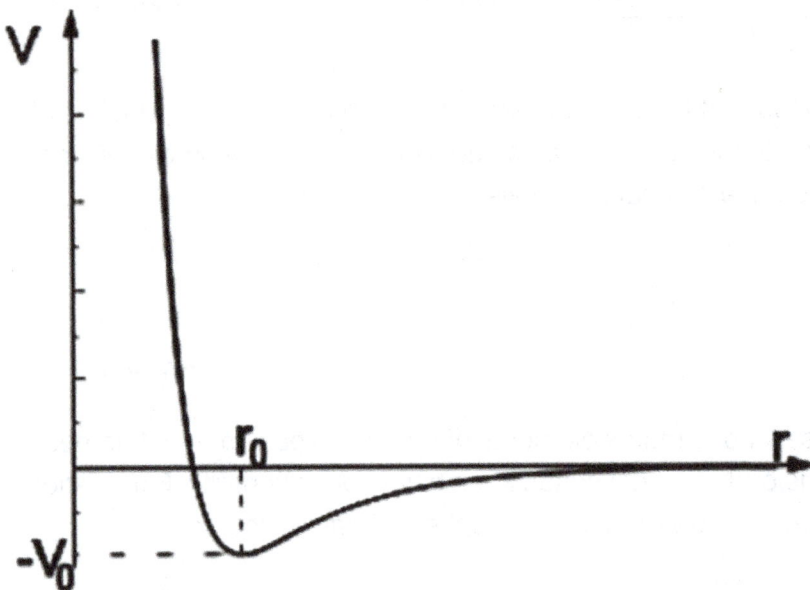

b. The key here is to approximate the Lennard-Jones potential as a harmonic potential for small motion, then use the formula for the harmonic oscillator energies. Want to expand about the internuclear distance at which the first derivative is 0:

$$\frac{dV}{dr} = \frac{V_0}{r} \left[-12 \left(\frac{r_0}{r} \right)^{12} + 12 \left(\frac{r_0}{r} \right)^6 \right]$$

$$\left.\frac{dV}{dr}\right|_{r=r_{min}} = \frac{V_0}{r_{min}}\left[-12\left(\frac{r_0}{r_{min}}\right)^{12} + 12\left(\frac{r_0}{r_{min}}\right)^6\right] = 0$$

$r_{min} = r_0$

Now get the harmonic approximation from the second derivative:

$$\frac{d^2V}{dr^2} = \frac{V_0}{r^2}\left[(12)(13)\left(\frac{r_0}{r}\right)^{12} - (12)(7)\left(\frac{r_0}{r}\right)^6\right]$$

$$\left.\frac{d^2V}{dr^2}\right|_{r=r_{min}} = \frac{V_0}{r^2}\left[(12)(13) - (12)(7)\right] = \frac{72V_0}{r_0^2}$$

So the potential is harmonic with the spring constant given by the value of the second derivative at minimum. So the energy levels are:

$$\hbar\left(n + \frac{1}{2}\right)\sqrt{\frac{k}{\mu}} = \hbar\left(n + \frac{1}{2}\right)\sqrt{\frac{72V_0}{r_0^2\mu}} = \hbar\left(n + \frac{1}{2}\right)\sqrt{\frac{144V_0}{r_0^2 m}} = 12\hbar\left(n + \frac{1}{2}\right)\sqrt{\frac{72V_0}{r_0^2 m}}$$

The reduced mass for identical particles is simply:

$$\mu = \frac{m_1 m_2}{m_1 + m_2} = \frac{m}{2}$$

c. The rotational energies are given by the angular momentum operator:

$$H\Psi = \frac{L^2}{2I}\Psi = \frac{\hbar^2 L(L+1)}{2I}\Psi$$

I is the moment of inertia of the diatomic molecule:

$$I = \mu r^2 = \frac{mr_0^2}{2} \text{ (for identical molecules separated by } r_0)$$

The vibrational transitions will be in the IR, and the rotational transitions in the microwave region.

Solution 4.28

From Problem 4.27, these energies correspond to:

$$\frac{\hbar^2}{2I}J(J+1) \text{ for } J = 9, 8, 7, 6, 5, 4, 3$$

Use this to get the moment of inertia I as $I = 2.71 \times 10^{-40}$ g cm². Then the separation distance comes from the moment of inertia for a rigid rotor $I = \mu R^2$, giving $d = 1.3 \times 10^{-8}$ cm.

Solution 4.29

a. For large separation, it is energetically favorable to have the electron on the K. The trade-off point is when that gain in energy is balanced by the Coulomb interaction:

$$\frac{e^2}{R} = 4.32 - 3.80 \text{eV} = 0.52 \text{ eV}$$

$$R = 27.7 \text{ Å}$$

b. The trick here is to set the highest J state (J_{max}) to the energy at which the molecule becomes unbound—that is, to 0.52 eV. Knowing the equilibrium separation, we can get the moment of inertia I.

$$\frac{\hbar^2}{2I}J_{max}^2 = 0.52 \text{ eV}$$

$$I = \frac{m_1 m_2}{m_1 + m_2} r_{eq}^2$$

$$J_{max} \sim 180$$

Solution 4.30

a. For the phase shifts, have the expression:

$$\tan\delta_\ell = \frac{kj_\ell'(ka) - \gamma_\ell j_\ell(ka)}{kj_\ell'(na) - \gamma_\ell n_\ell(ka)}$$

Take:

$$j_\ell(r) \xrightarrow{r\to 0} \frac{r^\ell}{(2\ell+1)!!}$$

$$n_\ell(r) \xrightarrow{r\to 0} \frac{(2\ell-1)!!}{r^{\ell+1}}$$

Now use the first three partial waves:

$$j_0(\rho) \to 1, j_1(\rho) \to \frac{\rho}{3}, j_2(\rho) \to \frac{\rho^2}{15}$$

$$n_0(\rho) \to -\frac{1}{\rho}, n_1(\rho) \to -\frac{1}{\rho^2}, n_2(\rho) \to -\frac{3}{\rho^3}$$

$$j_0'(\rho) = -j_1(\rho), j_\ell' = j_{\ell-1} - \frac{\ell+1}{\rho}j_\ell$$

So:

$$j_0' \to -\frac{\rho}{3}$$

$$j_1' \to j_0 - \frac{2}{\rho}j_1 = \frac{1}{3}$$

$$j_2' \to j_1 - \frac{3}{\rho}j_2 = \frac{2}{15}\rho$$

$$n_0' \to \frac{1}{\rho^2}$$

$$j_1' \to -\frac{1}{\rho} + \frac{2}{\rho^3}$$

$$j_2' \to -\frac{1}{\rho^2} - \frac{9}{\rho^4}$$

Gives:

$$\tan\delta_0 \approx \frac{k\left(\frac{-ka}{3}\right) - \gamma_0(1)}{\frac{1}{(ka)^2}k + \frac{\gamma_0}{ka}} \approx \frac{-ka^2\gamma_0}{1 + \gamma_0 a}$$

$$\tan\delta_1 \approx \frac{\frac{k}{3} - \gamma_1\left(\frac{ka}{3}\right)}{k\left(-\frac{1}{ka} + \frac{2}{(ka)^3}\right) + \frac{\gamma_1}{(ka)^2}} = \frac{(ka)^3}{3}\left(\frac{1 - a\gamma_1}{2 + a\gamma_2}\right)$$

Similarly:

$$\tan\delta_2 \approx -\frac{k^5 a^5}{15}\left(\frac{2 - a\gamma_2}{9 - 3a\gamma_2}\right)$$

The total cross-section is given by:

$$\sigma = \frac{4\pi}{k^2}\sum_\ell (2\ell + 1)\sin^2\delta_\ell = \frac{4\pi}{k^2}\left[3\delta_1^2 + 5\delta_2^2\right]$$

For the very first term, we can look at the gammas and approximate:

$$\delta_1 \approx -\frac{(ka)^3}{3}$$

Giving:

$$\sigma \approx \frac{4}{3}\pi a^6 k^4$$

The differential cross-section is given by:

$$\frac{d\sigma}{d\Omega} = \frac{1}{k^2}\left|\sum_{\ell=1}^{\infty}(2\ell + 1)\exp(i\delta_\ell)\sin\delta_\ell P_\ell(\cos\theta)\right|^2$$

For $\ell = 1, 2$:

$$\frac{d\sigma}{d\Omega} = \frac{1}{k^2}\left|3\exp(i\delta_1)\sin\delta_1\cos\theta + 5\exp(i\delta_2)\sin\delta_2\frac{1}{2}(3\cos^2\theta - 1)\right|^2$$

$$= \frac{1}{k^2}\left[9\sin^2\delta_1\cos^2\theta + \frac{25}{4}\sin^2\delta_2\frac{1}{2}(3\cos^2\theta - 1)^2 + 15\sin\delta_1\sin\delta_2\cos(\delta_1 - \delta_2)\cos\theta(3\cos^2\theta - 1)\right]$$

Now use the approximations:

$$\sin^2\delta_{1,2} \approx \delta_{1,2}$$

$$\cos(\delta_1 - \delta_2) \approx 1$$

Giving:

$$\frac{d\sigma}{d\Omega} \approx \frac{1}{k^2}\left[9\delta_1\cos^2\theta + \frac{25}{4}\sin^2\delta_2\frac{1}{2}(3\cos^2\theta - 1)^2 + 15\delta_1\delta_2\cos\theta(3\cos^2\theta - 1)\right]$$

The leading term is just:

$$\frac{1}{k^2}\left[9\delta_1\cos^2\theta\right] = k^4 a^6\cos^2\theta$$

b. We want to have:

$$\delta_0 = \pi$$

$$\tan^{-1}\left(-ka + \frac{k}{\alpha}\tan\alpha a\right) = \pi$$

Solve numerically for the minimum; get $\alpha = 4.49$. This gives:

$$\alpha a = \sqrt{\frac{2m(E + V_0)}{\hbar^2}}\,a = 4.49$$

$$V_0 a^2 \approx 10\frac{\hbar^2}{m}$$

Solution 4.31

In the semi-classical approximation (large ℓ), the WKB approximation may be used. This gives a general solution for the phase shift created by any potential $V(r)$:

$$\delta_\ell = \frac{mV(r_0)r_0}{k\hbar^2}, \text{ where } r_0 \sim \frac{\ell}{k}$$

In this case, $\delta_\ell = \frac{mA}{\hbar^2\ell}$.

For small ℓ, this is not an appropriate approximation. The Born approximation can be used instead of the WKB approximation in this case:

$$\chi'' + \left[k^2 - \frac{\ell(\ell+1)}{r^2} - \frac{2m}{\hbar^2}V \right]\chi = 0$$

$$\chi_0'' + \left[k^2 - \frac{\ell(\ell+1)}{r^2} \right]\chi_0 = 0$$

The assumption here is that V is a perturbation, so $\chi \sim \chi_0$. Multiply both sides by χ_0 to give:

$$\chi'(r)\chi_0(r) - \chi_0'(r)\chi(r) = \frac{2m}{\hbar^2}\int_\infty^0 V\chi\chi_0 dr$$

$$R_\ell = 2rj_\ell(\alpha r)$$

$$\xrightarrow{r\to\infty} \frac{2\sin\left(\alpha r - \frac{\ell\pi}{2}\right)}{r} \text{ (unperturbed)}$$

$$\xrightarrow{r\to\infty} \frac{2\sin\left(\alpha r - \frac{\ell\pi}{2} + \delta_\ell\right)}{r} \text{ (perturbed)}$$

$$\Rightarrow \sin\delta_\ell = \frac{-\pi m}{\hbar^2}\int_\infty^0 V(r)\left[j_\ell(\alpha r)\right]^2 dr = \frac{-\pi m}{\hbar^2}\int_0^\infty \frac{A}{r}\left[j_\ell(\alpha r)\right]^2 dr$$

$$\delta_\ell \approx \frac{-\pi m A}{\hbar^2} \int\limits_0^\infty \frac{1}{r} \left[J_{\ell+\frac{1}{2}}(\alpha r) \right]^2 dr$$

For $A < 0$, δ_ℓ becomes imaginary— so at some point the solution decays exponentially, meaning that there is no bound state.

Solution 4.32

Problem 4.32

a. $A\psi = a\psi$

$A^2\psi = A\psi = aA\psi = a\psi$

$\Rightarrow a^2 = a$

$a = 0, 1$

b. $P^2 = 1$

So $(\frac{1}{2})(1 \pm P)$ satisfies the equation.

c. $(\frac{1}{2})(1 \pm P)\psi$ is the positive-parity part of the wave function, and $(\frac{1}{2})(1 - P)\psi$ is the negative parity part.

Solution 4.33

Problem 4.33

a. See diagram on following page.

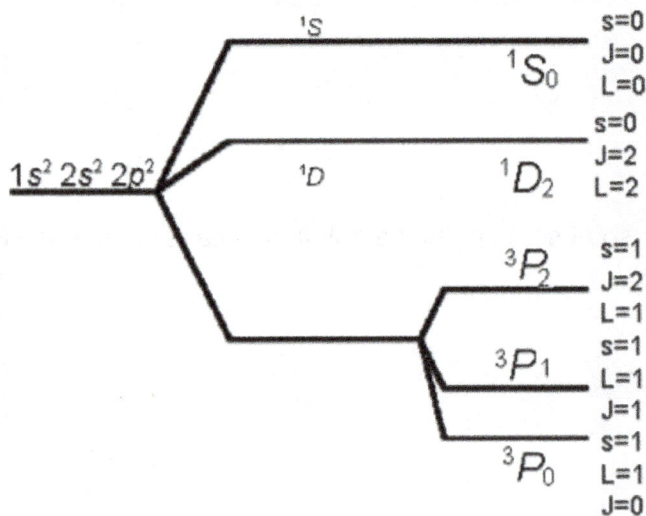

$$1s^2\,2s^2\,2p^2$$

States shown: ^1S, $^1S_0$ ($s=0$, $J=0$, $L=0$); ^1D, $^1D_2$ ($s=0$, $J=2$, $L=2$); $^3P_2$ ($s=1$, $J=2$, $L=1$); $^3P_1$ ($s=1$, $L=1$, $J=1$); $^3P_0$ ($s=1$, $L=1$, $J=0$)

b. $\quad L_z = \dfrac{\left(\vec{J}\cdot\vec{J} + \vec{L}\cdot\vec{L} - \vec{S}\cdot\vec{S}\right) J_z}{2J(J+1)}$

$\quad S_z = \dfrac{\left(\vec{J}\cdot\vec{J} - \vec{L}\cdot\vec{L} + \vec{S}\cdot\vec{S}\right) J_z}{2J(J+1)}$

$\quad \mu_{LSJM} = \left\langle \left[\varphi^L,\chi^S\right]_M^J \middle| g_L L_z + g_S S_z \middle| \left[\varphi^L,\chi^S\right]_M^J \right\rangle$

$\quad\quad = M\hbar\left[\dfrac{g_L + g_S}{2} + \dfrac{g_L - g_S}{2}\dfrac{L(L+1) - S(S+1)}{J(J+1)}\right]$

$\quad\quad = 0$ for $^1S_0$

$\quad\quad = g_L M\hbar$ for $^1D_2$

$\quad\quad = \dfrac{g_L + g_S}{2} M\hbar$ for ^3P states

Solution 4.34

$\pm 2\hbar$, with equal probability

Solution 4.35

Problem 4.35

a. See Problem 4.21. Because electrons repel each other via a Coulomb repulsion, the space-symmetric state (triplet) because the entire wave function must be antisymmetric.

b. There will be a spatial wave function only, no spin wave function. So only the space-symmetric levels will exist; the antisymmetric levels (corresponding to the spin-triplet states in normal He) will be absent.

Solution 4.36

Problem 4.36

Larmor derived the formula for the radiation from a non-relativistic charge as it accelerates back in 1897. We can calculate the acceleration due to transitions as follows:

$$a = \frac{F}{m} = \frac{e^2}{mr^2}$$

$$r = \frac{n^2\hbar^2}{me^2}$$

Now plug into the Larmor formula to give:

$$P = \frac{2}{3}\frac{m^2 e^{14}}{c^3 \hbar^8 n^8}$$

Solution 4.37

Problem 4.37

$$F \sim \frac{1\text{ eV}}{\text{Å}}\left(\frac{10^{16}\text{ atoms}}{\text{cm}^2}\right)\left(1.6\times10^{-12}\frac{\text{erg}}{\text{eV}}\right)\left(\frac{\text{Å}}{10^{-8}\text{ cm}}\right) = 1.6\times10^{12}\frac{\text{dyne}}{\text{cm}^2}$$

Solution 4.38

a. $\Delta V = hc \dfrac{\Delta\lambda}{\lambda_1\lambda_2} = 0.785\ \text{V}$

b. Work function is given by $W = h\nu - K$, where K is the kinetic energy of the ejected electron. K is also related to the stopping potential by $K = qV$. So plug in:

$$W = \frac{1240\ \text{eVnm}}{545\ \text{nm}} - 1.07\ \text{eV} = 1.2\ \text{eV}$$

Solution 4.39

Use Pauli spin matrices:

$$s_y = \begin{pmatrix} 0 & -i \\ i & 0 \end{pmatrix}$$

$$s_z = \begin{pmatrix} 1 & 0 \\ 0 & -1 \end{pmatrix}$$

Solve for eigenvalues λ:

$$\det\begin{pmatrix} B - \lambda & -iA \\ iA & -B - \lambda \end{pmatrix} = 0$$

$$\lambda^2 = A^2 + B^2$$

So the eigenvalues are $\pm\dfrac{\hbar}{2}\sqrt{A^2 + B^2}$.

Now construct normalized eigenvectors from these eigenvalues. For a general 2×2, the elements of the eigenvector v_1 and v_2 can be calculated readily to give:

$$v_2 = \frac{\lambda - a_{11}}{a_{12}} = \frac{-B \pm \sqrt{A^2 + B^2}}{-iA}, v_1 = 1$$

Then normalizing:

$$v_1 = \frac{1}{\sqrt{1 + v_2^2}}, \quad v_2 = \frac{v_2}{\sqrt{1 + v_2^2}},$$

which gives the eigenvector:

$$\psi = \frac{1}{\sqrt{2B\left(B \pm \sqrt{A^2 + B^2}\right)}} \begin{bmatrix} iA \\ B \mp \sqrt{A^2 + B^2} \end{bmatrix}$$

Solution 4.40

Problem 4.40

Assume that the electron and neutrino depart immediately (the "sudden approximation"). Then the initial and final states are simply those of a hydrogenic atom of charge Z and $Z + 1$, respectively. The probability of being in the ground state is then given by:

$$P = \left| \langle \psi_0(Z + 1) | \psi_0(Z) \rangle \right|^2$$

The ground state of hydrogenic atoms is given by:

$$\psi_0 = \frac{1}{\sqrt{\pi}} \left(\frac{Z}{a_0} \right)^{\frac{3}{2}} \exp\left(\frac{-Zr}{a_0} \right)$$

So all that remains to do is to solve the integral:

$$P = \frac{1}{\pi} \left(\frac{Z(Z + 1)}{a_0^2} \right)^3 \left| \int \exp\left(-[2Z + 1]r/a_0 \right) d^3r \right|^2$$

$$= 16 \left(\frac{Z(Z + 1)}{a_0^2} \right)^3 \left| \int \exp\left(-[2Z + 1]r/a_0 \right) r^2 dr \right|^2$$

Using the integral: $\displaystyle\int_0^\infty e^{-r} r^2 dr = 2$

Gives $P = 64 \left[\dfrac{Z(Z+1)}{(2Z+1)^2} \right]^3$

Solution 4.41

Problem 4.41

The Born approximation assumes that the potential is a perturbation that does not significantly affect the wave function, so:

$$\frac{d\sigma}{d\Omega} = \left| f\left(\vec{k}',\vec{k}\right) \right|^2$$

$$f\left(\vec{k}',\vec{k}\right) \approx \frac{-m}{2\pi\hbar^2} \int \exp\left[i\left(\vec{k} - \vec{k}'\right) \right] \cdot \vec{r}' V\left(\vec{r}'\right) d^3\vec{r}'$$

For a spherically symmetric potential:

$$f\left(\vec{k}',\vec{k}\right) \approx \frac{-m}{2\pi\hbar^2} \iiint \exp\left(iqr'\cos\theta' \right) \cdot r'^2 V(r') dr' \sin\theta' d\theta' d\varphi'$$

$$= \frac{-2m}{\hbar^2 q} \int r' V(r') dr' \sin qr' dr', \text{ where } q \equiv \left| \vec{k} - \vec{k}' \right| = 2k\sin\left(\frac{\theta}{2}\right)$$

In the case of the Yukawa potential, this becomes:

$$f\left(\vec{k}',\vec{k}\right) \cong \frac{2mg^2}{\hbar^2 q} \int \exp(-r'/R) dr' \sin qr' dr' = \frac{2mg^2}{\hbar^2} \frac{1}{q^2 + \left(\frac{1}{R}\right)^2}$$

Giving:

$$\frac{d\sigma}{d\Omega} = \left(\frac{2mg^2}{\hbar^2} \right)^2 \frac{1}{\left[q^2 + \left(\frac{1}{R}\right)^2 \right]^2},$$

where m is the reduced mass of the neutron and proton, approximately equal to $\dfrac{m_n}{2}$, where m_n represents a general nucleon mass.

Solution 4.42

a. The particle will be a traveling wave outside the barrier and a decaying exponential inside the barrier. Write the equations for before, inside, and after in terms of 4 constants A, B, C, and D.

For $x < 0$:

$$\psi = \exp(ikx) + A\exp(-ikx)$$

$$\frac{d\psi}{dx} = ik\exp(ikx) - A(ik)\exp(-ikx)$$

For $0 < x < a$:

$$\psi = B\exp(k'x) + C\exp(k'x)$$

$$\frac{d\psi}{dx} = Bk'\exp(k'x) - Ck'\exp(-k'x)$$

For $x > a$:

$$\psi = D\exp(ikx)$$

$$\frac{d\psi}{dx} = D(ik)\exp(ikx)$$

Where $k = \sqrt{\dfrac{2mE}{\hbar^2}}$, $k' = \sqrt{\dfrac{2m(V_0 - E)}{\hbar^2}}$.

In order to determine the coefficients, set the equations and derivatives equal at the boundary conditions.

For $x = 0$:

$$1 + A = B + C$$

$$ik - A(ik) = Bk' - Ck'$$

For $x = a$:

$$B\exp(k'a) + C\exp(-k'a) = D\exp(ika)$$

$$Bk'\exp(k'a) - Ck'\exp(-k'a) = D(ik)\exp(ika)$$

Solve these equations for B, and get transmission probability:

$$P = (B)^2 = \frac{4k^2k'^2}{\left(k^2 + k'^2\right)^2\sinh^2(k'a) + 4k^2k'^2}.$$

This $\to 0$ as $\hbar \to 0$, as would be expected as it's a purely quantum mechanical phenomenon.

b. For the proton, 3.9×10^{-19}. For the electron, 0.779. Definitely this makes sense, since the proton is heavier and thus more classical. Tunneling is seen and used on a regular basis with light particles such as electrons, but is rarer for heavier particles, causing events such as radioactive decay.

Solution 4.43

Problem 4.43

Treat the system as a particle in a box. Then the zero-point energy is:

$$\frac{h^2}{8mL}$$

Even if we made L very large (say 1 meter), the energy is still on the order of 10^{-67} J.

Solution 4.44

Problem 4.44

The electron is moving in the field of its image charge. The Schrodinger equation is just that of the hydrogen atom, except with e^2 replaced by $\frac{e^2}{2}$. The energy to remove it is thus just the ionization energy, $\frac{me^4}{4\hbar^2}$.

Solution 4.45

Problem 4.45

The Hamiltonian is given by:

$$H = -\frac{\hbar^2}{2Nmr^2}\frac{d^2}{d\theta^2},\text{ which has solutions of the form:}$$

$$\psi(0) = \exp(iNk\theta),\ E_k = \frac{\hbar^2 Nk^2}{2mr^2},\text{ with degenerate } k \text{ states } 0, \pm1, \text{ etc.}$$

In the limit of a ring, the energy difference between the ground state and first excited state goes to infinity.

Solution 4.46

Problem 4.46

a. If $\Im(I)$ changes slowly with I, then we can assume neighboring levels have the same $\Im$, that is:

$$E(I) - E(I-2) \approx \frac{\hbar^2}{2\Im}\left[I(I+1) - (I-1)(I-2)\right],\text{ which gives}$$

$$\frac{2\Im}{\hbar^2} \approx \frac{4I-2}{E(I) - E(I-2)}$$

b. To get the angular velocity, use: $\Im\omega = \hbar\sqrt{I(I+1)}$, expressed as

$$(\hbar\omega)^2 = I(I+1)\left(\frac{\hbar^2}{\Im}\right)^2.\text{ Now plug in the solution from the first part to obtain}$$

$$(\hbar\omega)^2 \approx 4I(I+1)\left(\frac{E(I) - E(I-2)}{4I-2}\right)^2.$$

c. Looking at the plot, it can be seen that there is a dramatic change in the curve at about $\frac{2\Im}{\hbar^2} = 0.9$, indicating a critical point where the structure of the nucleus is altered. The moment of inertia approaches that of a rigid rotor, suggesting that what is happening is analogous to the superconductor-normal transition in a metal: as the nucleus rotates faster and faster, more unpaired particles fill the levels near the Fermi surface. This causes the pair correlation parameter to vanish, leaving a rigid rotator.

Solution 4.47

a. Term in a: SHIFT because p^4 has a non-vanishing expectation value, but is diagonal and independent of the spin states. Term in b: Neither, since the 1s level has angular momentum of zero and therefore all actions of the L operator bring a zero. Term in c: This causes a shift, but not a splitting, since the 1s electron has a certain probability of being at the nucleus which is not a function of the electron's spin and only one angular momentum state exists at this level. Term in d: Neither; as in part (b), the L operator is zero for the 1s state. Term in e: The matrix

$$\left\langle m_e, m_p \left| \vec{S} \cdot \vec{I} \right| m_e, m_p \right\rangle \approx \left\langle m_e, m_p \left| S_z I_z + \frac{1}{4}(S_+ + S_-)(I_+ + I_-) + \frac{1}{4i}(S_+ - S_-)(I_+ - I_-) \right| m_e, m_p \right\rangle$$

$$L_+ | - \rangle = \hbar | + \rangle \quad L_- | + \rangle = \hbar | - \rangle \quad L_z | \pm \rangle = \frac{\hbar}{2} | \pm \rangle$$

is non-diagonal and so introduces a splitting and a shift. Further, the other term is possibly diagonal or possibly non-diagonal depending on R. Term in f: Just as in term (c), a constant effect is brought by the delta function. The matrix

$$\left\langle m_e, m_p \left| \vec{S} \cdot \vec{I} \right| m_e, m_p \right\rangle$$ is non-diagonal and so introduces a splitting and a shift.

b. First, note that:

$$\left(\vec{S} + \vec{I} \right)^2 = \vec{S}^2 + \vec{I}^2 + 2\vec{S} \cdot \vec{I}$$

$$\vec{S} \cdot \vec{I} = \frac{1}{2}\left(\vec{F}^2 - \vec{S}^2 - \vec{I}^2 \right)$$

As seen in part (a), the d term is zero, so only have to consider the other ones:

$$\left\langle \psi \left| \left\langle F', m'_F \left| \frac{e}{R^3} \frac{3\left(\vec{S} \cdot \vec{R} \right)\left(\vec{I} \cdot \vec{R} \right)}{R^2} - \vec{S} \cdot \vec{I} + f\left(\vec{S} \cdot \vec{I} \right)\delta(R) \right| F, m_F \right\rangle \right| \psi \right\rangle$$

The only elements that survive are:

$$\left\langle \psi \left| H_{hf} \right| \psi \right\rangle | F, m_F \rangle = \vec{S} \cdot \vec{I} \left| F, m_F \right\rangle \left[e\left\langle \psi \left| \frac{x^2}{R^5} \right| \psi \right\rangle + e\left\langle \psi \left| \frac{1}{R^3} \right| \psi \right\rangle + f\left\langle \psi \left| \delta(R) \right| \psi \right\rangle \right]$$

References

Quantum mechanics

Original references to some of the problems in the chapter:

1. Geiger, H. and Nuttall, J.M. The ranges of the α particles from various radioactivesubstances and a relation between range and period of transformation. *Philosophical Magazine* **22**, 613-621 (1911).

2. Aharonov, Y. and Bohm, D. Significance of electromagnetic potentials in quantum theory. *Physical Review* **115**, 485–491 (1959).

3. Breit, G. Theory of Isotope Shift. *Reviews of Modern Physics* **30**, 507-517 (1958).

4. Cashion, J.K. Simple Formulas for the Vibrational and Rotational Eigenvalues of the Lennard-Jones 12-6 Potential. *J Chem Phys* **48**, 94-103 (1968).

5. Deck, T.J.G., Amar, J.G. and Fralick, G. Nuclear size corrections to the energy levels of single-electron and -muon atoms. *J. Phys. B: At. Mol. Opt. Phys.* **38**, 2173–2186 (2005).

6. Gamow, G. Zur Quantentheorie des Atomkernes (On the quantum theory of the atomic nucleus). *Zeitschrift für Physik* **51**, 204-212 (1928).

7. Lamb, W.E. and Retherford, R.C. Fine structure of the hydrogen atom by a microwave method. *Physical Review* **72**, 241-243 (1947).

8. Mossbauer, R. Recoilless Nuclear Resonance Absorption. *Annual Review of Nuclear Science* **12**, 123-152 (1962).

9. Tucker, T.C., Roberts, L.D., Nestor, C.W. and Carlson, T.A. Calculation of the Electron Binding Energies and X-Ray Energies for the Superheavy Elements 114, 126, and 140 Using Relativistic Self-Consistent- Field Atomic Wave Functions. *Phys Rev* **174**, 118-124 (1968).

10. Crommie, M.F., Lutz, C.P. and Eigler, D.M. Confinement of electrons to quantum corrals on a metal surface. *Science* **262**, 218-220 (1993).

11. Born, M. and Oppenheimer, J.R. Zur Quantentheorie der Molekeln [On the Quantum Theory of Molecules]. *Annalen der Physik* **389**, 457-484 (1927).

12. Larmor, J. On a dynamical theory of the electric and luminiferous medium. *Philosophical Transactions of the Royal Society* **190**, 205-300 (1897).

Also see:

13. Griffiths, D.J. *Introduction to Quantum Mechanics*, (Prentice Hall, Saddle River, NJ, 2004).

14. Robinett, R.W. Visualizing the solutions for the circular infinte well in quantum and classical mechanics. *American Journal of Physics* **64**, 440-446 (1996).

15. Bailey, V.A. and Townsend, J.S. The motion of electrons in gases. *Philosophical Magazine* **S.6**, 873-891 (1921).

16. Bohm, D. *Quantum Theory*, (Prentice-Hall, Englewood Cliffs, New Jersey, 1951).

17. Schiff, L.I. *Quantum Mechanics*, (McGraw-Hill, New York, 1968).

18. Sakurai, J.J. *Modern Quantum Mechanics*, (Addison-Wesley, Redwood City, CA, 1985).

19. Messiah, A. *Quantum Mechanics*, (Dover Publications, Mineola, NY, 1999).

5

Statistical and thermal physics

Problems
Statistical and thermal physics

Problem 5.01

Solution 5.01

a. Give an expression for the entropy for a series of N measurements with probability p_i each. For what values of p_i is entropy maximized?

b. Take as an example two series of four measurements, one with probabilities 0.25, 0.25, 0.25, 0.25 and the second with 0.75, 0.2, 0.05, 0. Which has the higher entropy? What does this mean?

Problem 5.02

Solution 5.02

Prove the Stirling approximation:

$$n! \approx \sqrt{2\pi n}\, n^n e^{-n}$$

Problem 5.03

Solution 5.03

For a perfect gas in the Grand Canonical ensemble, find the probability of having N particles in a given volume V.

Problem 5.04

Solution 5.04

Find the density of states, Fermi energy, and chemical potential for a 2-dimensional Fermi gas.

Problem 5.05

Solution 5.05

Two gases are separated in a cylindrical volume by a freely sliding piston. Particles of mass m are placed in one side, and particles of mass $2m$ in the other side. Calculate the equilibrium number densities in the limits of very high and very low temperatures, assuming:

a. The particles are fermions with spins 1/2 and 3/2 respectively.

b. The particles are spinless bosons.

Problem 5.06

Solution 5.06

Find the Fermi energy of an ultra-relativistic Fermi gas in 3D.

Problem 5.07

Solution 5.07

Can Bose condensation occur in an ultra-relativistic Bose gas of arbitrary dimension? If so, give an expression for the critical temperature T_c.

Problem 5.08

Solution 5.08

In general, for a d-dimensional gas of bosons with energy-momentum relation $\varepsilon \sim p^x$, when can Bose condensation occur?

Problem 5.09

Solution 5.09

Calculate the speed of sound in a gas at $T = 0$.

Problem 5.10

Solution 5.10

Find an expression for the entropy per particle of a free Bose/Fermi gas (no need to solve the integral). Show that this implies that in an adiabatic process, $VT^{3/2}$ and $PV^{5/3}$ are constant.

Problem 5.11

Solution 5.11

A classical gas of noninteracting dipoles (each of moment μ) is placed into a uniform electric field. Derive an expression for the polarization.

Problem 5.12

Solution 5.12

Give the statistical thermodynamic derivation of the van der Waals equation.

Problem 5.13

Solution 5.13

Show that the specific heat capacity C_V of a van der Waals gas is independent of volume.

Problem 5.14

Solution 5.14

Expand the equation of state for a van der Waals gas around its critical variables T_c and n_c. Show that the isothermal compressibility diverges and that $n - n_c$ vanishes as the temperature approaches T_c.

Problem 5.15

Solution 5.15

Take the classical limit of the expression for the thermodynamic potential of a Bose/Fermi gas and show that it gives the ideal gas equation.

Problem 5.16

Solution 5.16

At the helium λ point, how does T_λ compare to the Bose condensation transition temperature? Is the superfluid transition the same thing as Bose condensation? Discuss.

Problem 5.17

Solution 5.17

Find the temperature for which $\mu = 0$ for a Fermi gas. Why does "Fermi condensation" not take place?

Problem 5.18

Solution 5.18

Define an "ergodic system." Under what circumstances is a system of two uncoupled harmonic oscillators ergodic?

Problem 5.19

Solution 5.19

Define the quantum Hall effect and explain how it works in a 2D electron gas.

Problem 5.20

Solution 5.20

The white dwarf mass-radius condition is the limit of existence of a white dwarf without collapsing: that is, the point at which the gravitational pressure inwards equals the outwards degeneracy pressure. The star may be modeled as a core of He nuclei mingling with a degenerate electron gas.

a. Assuming degeneracy of the fermions $(\beta\mu \gg 1)$, calculate the outward pressure.

b. Assuming non-degeneracy of the core nuclei $(\beta\mu \ll 1)$, calculate the gravitational pressure. Use the mean-field gravitational potential $W = -GMm_{\mathrm{He}}/R$, where M and R are the mass and radius of the star, respectively.

c. Equate the two to give the non-relativistic mass-radius condition. This is called the Chandrasekhar limit.

Problem 5.21

Solution 5.21

Below a frequency called the plasma frequency, electromagnetic waves cannot propagate through ionized gas. Derive a formula for the plasma frequency for a completely ionized hydrogen gas.

Problem 5.22

Solution 5.22

Using free electron theory, the resistivity of a metal is proportional to temperature to some power. Show that this power is 3/2 classically and 1 quantum mechanically. You may assume that the mean free path λ is proportional to $1/T$.

Problem 5.23

Solution 5.23

Doping of semiconductors allows for energy states within the bandgap; donor impurities are those with energies near the conduction band. If a semiconductor contains N_d donor impurities, express N_d as a function of the Fermi energy, donor level energy, and energy of the conduction band.

Problem 5.24

Solution 5.24

The partition function for a diatomic molecule can be written as a function of three energy parameters–electronic *el*, vibrational *vib*, and rotational *rot*–as follows:

$$Z = \frac{e^{\beta \varepsilon_{el}}}{1 - e^{\beta \varepsilon_{vib}}} \sum_{\ell} (2\ell + 1) e^{-\beta \ell (\ell + 1) \varepsilon_{rot}}$$

a. Write an expression for the vibrational partition function and calculate the average vibrational energy.

b. In the low-temperature limit, only the first two terms in the rotational partition function are needed. What are these terms for (i) a molecule with non-identical atoms, and (ii) a molecule consisting of identical spinless bosons?

c. What is the average rotational energy in the high-temperature limit?

Problem 5.25

Solution 5.25

The moon's radius is 1738 km, its mean temperature is 300 K, and its surface gravity is 1/6 that of Earth's. What molecular weight must a gas have so that its RMS velocity will not allow it to escape the moon? How does this explain the lack of atmosphere on the moon?

Problem 5.26

Solution 5.26

A gas of two types of nonidentical spin-½ fermions is placed in a magnetic field. The magnetic moments of the two particle types are:

$$\mu_1 = \frac{g_1\mu}{2} > 0$$

$$\mu_2 = \frac{g_2\mu}{2} < 0$$

Calculate the net spin in the direction of the field in the high temperature limit and low temperature limit. What condition would allow the net spin to be parallel to the field at low temperatures and anti-parallel at high temperatures?

Problem 5.27

Solution 5.27

A polymer is made up of N subunits each of length x. Each link is equally likely to point right or left.

a. How many different arrangements yield a length $l = 2mx$, where m is an integer?

b. Given an expression for the entropy to first order in l assuming large N.

c. What is the free energy?

d. Assuming an ideal polymer, what is the retractive force?

Problem 5.28

Solution 5.28

A solid consists of N non-interacting spin-1 particles. Possible quantum states for each particle are $m = 0, \pm 1$, where the $m = 0$ state has energy $= 0$ and the $m = \pm 1$ states both have energy $= \varepsilon > 0$. Calculate the total number of states with $E = n\varepsilon$, and use this to derive an expression for the entropy. Use this formula to derive an expression for the temperature as a function of ε; explain when $T < 0$ and what that means.

Problem 5.29

Solution 5.29

Derive the Clausius-Clapeyron equation that relates vapor pressure P to temperature in terms of the latent heat of vaporization L and the volume change between liquid and gas ΔV:

$$\frac{dP}{dT} = \frac{L}{T\Delta V}$$

Problem 5.30

Solution 5.30

If paper burns at 451 °F (233 °C) in the world as we know it, how would this change in a universe where mass of the electron was doubled?

Problem 5.31

Solution 5.31

Helium is an example of a gas that is well described by the ideal gas law. It boils at $T_0 = 4.2$ K with vapor pressure $P_0 = 1$ atm. If you pump on liquid helium and reduce the pressure to $P_f \ll P_0$, what is the temperature of the liquid?

Problem 5.32

Solution 5.32

Calculate the rms velocity of particles escaping through a small hole in a container (effusion), and compare with that of the rms velocity of the particles inside the container.

Problem 5.33

Solution 5.33

A building is heated by using a Carnot engine running backwards, extracting heat at a well at T_1 and delivering it to a reservoir at $T_2 > T_1$. How much energy is needed to drive the heat pump if the total energy delivered to a reservoir is Q?

Problem 5.34

Solution 5.34

Consider a magnetic system where the internal energy U and magnetization M are given by:

$$U = C_M T$$

$$M = \frac{aH}{T}$$

where H is the field strength, C_M is the (constant) specific heat at constant M, and a is a constant of proportionality. Calculate the specific heat at constant field (C_H) and the change in entropy ΔS resulting from a change in field.

Problem 5.35

Solution 5.35

Consider a cubic conducting cavity of volume $V = L^3$. Express the pressure as a function of temperature and volume. What happens to the temperature if the volume expands adiabatically to side length $2L$?

Problem 5.36

Solution 5.36

When an atom in a crystal moves from its usual lattice position to an interstitial site, this forms what is known as a Frenkel defect. In a monoatomic crystal with N atoms, M interstitial sites, and n atoms in interstitial positions, what is the entropy of the system?

What is the energy E_1 needed to move an atom from its usual position to an interstitial position? You can assume that $n \ll N, M$.

Problem 5.37
Solution 5.37

A pressurized tank of volume V contains a monoatomic ideal gas at pressure $P = 100$ atm and temperature T. A valve is opened to slowly fill a balloon, until only 1/10 of the original gas molecules remain in the tank. Assuming there is no heat exchange, calculate the change in entropy of the gas in the tank and the final temperature. Also calculate the entropy of the gas leaving the tank and the heat added as it leaves. Compare this with $T\Delta S$.

Problem 5.38
Solution 5.38

Some "young-earth creationists" believe that the Sun is not a nuclear furnace, but that it is heated by gravitational contraction. This was the model proposed in the 1800s before the discovery of radioactivity, and is known as the Kelvin-Helmholtz mechanism. Given that the Sun's mass is $M = 2 \times 10^{30}$ kg, its radius is $R = 7 \times 10^8$ m, luminosity is $L = 4 \times 10^{26}$ watts, how long could it shine at a constant value of L? Compare this with the time scale possible for nuclear fusion, in which 7 MeV is produced for each hydrogen fusion reaction. What radius would the sun have to have to make these timescales match?

Problem 5.39
Solution 5.39

Derive the Stefan-Boltzmann law, which gives the energy density spectrum of blackbody radiation.

Problem 5.40

Solution 5.40

Neptune is a distance $d = 4.5 \times 10^{12}$ m from the Sun. Its mass is $M = 10^{26}$ kg, radius $R_n = 2.5 \times 10^7$ m, albedo $A = 0.41$. If the Sun is a black body radiating at $T = 5800$ K with radius $R_S = 7 \times 10^8$ m, what gas phase elements can we expect to see on Neptune? Comment on what was observed by Voyager.

Problem 5.41

Solution 5.41

Traditional home heating is done by burning a fuel such as wood. Can improvement upon this result from using some of the heat of combustion to power an engine to drive a heat pump?

Problem 5.42

Solution 5.42

A vacuum system contains molecules in the gas phase as well as adsorbed molecules on the surface. For a cubic vacuum system 0.4 m on a side, what is the gas pressure at which there will be as many nitrogen molecules in a single adsorbed monolayer as in the gas phase?

Problem 5.43

Solution 5.43

Consider a hollow body at temperature T containing electromagnetic radiation in equilibrium with its walls. The energy spectrum of this so-called blackbody radiation can be measured by observing it through a negligibly small hole in the cavity wall. As argued by Planck, the energy levels of the radiation are quantized and are the same as those of a simple harmonic oscillator of angular frequency ω.

a. What are the quantized energy levels of a radiation mode of angular frequency ω? Use the Boltzmann distribution to find the thermal average number of photons for such a mode.

b. The density of modes in $\vec{k}$ space is constant. Thus for a cavity of volume V, the number of modes in a volume d^3k near a wave number $\vec{k}$ is just $2\frac{V}{(2\pi)^3}d^3k$. If c is the speed of light, how many modes are there having an angular frequency between ω and $\omega + d\omega$?

c. Express the total internal energy U for this radiation inside the cavity as an integral over ω. Discuss analytically for very high and very low ω how the integral depends on ω. Sketch the integrand as a function of ω for two different values of T.

d. Show that $U = aVT^4$, and give an integral expression for the constant a.

e. Use part (d) to compute the entropy S and pressure P as functions of V and T. Determine any constants that appear in terms of the symbols introduced above. Show that your results are thermodynamically consistent with the formula in part (d).

Problem 5.44   Solution 5.44

Consider N identical but distinguishable very weakly interacting five-state quantum subsystems with energy levels shown as in the figure.

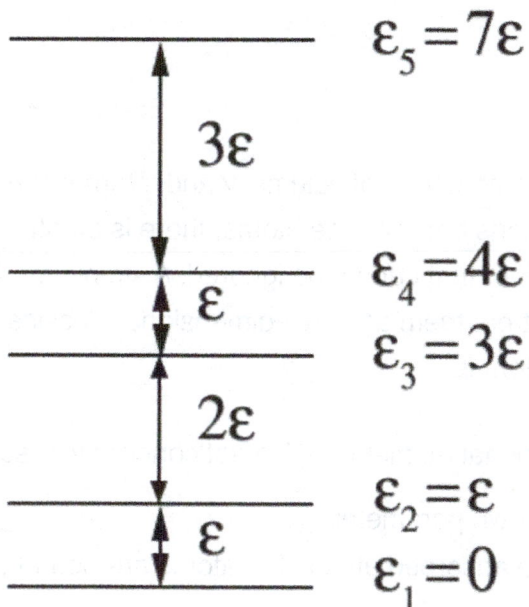

$$\varepsilon_5 = 7\varepsilon$$

3ε

$$\varepsilon_4 = 4\varepsilon$$
$$\varepsilon_3 = 3\varepsilon$$

ε

2ε

$$\varepsilon_2 = \varepsilon$$

ε

$$\varepsilon_1 = 0$$

a. What is the ratio of the fraction of systems in states ε_4 to those in states ε_2 for $k_B T = \frac{\varepsilon}{2}$?

b. How high, in units of $\frac{\varepsilon}{k_B}$, must be the temperature T be so that the minimum ratio of the population of any state to the population of any other state having a lower energy equals 0.99? In other words, what is the minimum temperature needed to guarantee that all states will be equally populated to with 1%?

c. Derive a formula for the internal energy U of the entire system of particles. For a very high temperature, say $T > 1000\frac{\varepsilon}{k_B}$, approximately what is the value of U? Give a numerical answer in units of $N\varepsilon$.

d. For $T = 0$, what is the entropy S of the entire system of particles? What is the justification for your answer?

e. For a very high temperature, say $T > 1000\frac{\varepsilon}{k_B}$, approximately what is the value of S?

f. Consider the heat capacity at constant volume, C_V, for the entire system of particles. Develop an approximate formula that is valid at very low temperatures, $T \ll \frac{\varepsilon}{k_B}$, and sketch the result vs. T. What is the asymptotic value of C_V for high temperatures, $T > 1000\frac{\varepsilon}{k_B}$?

Problem 5.45

Solution 5.45

Consider an ideal gas of N atoms confined to a container of volume V and internal surface area A. Although you may neglect the interactions between the atoms, there is an attraction between the atoms and the walls of the container that cannot be ignored. A simple model for the atoms adsorbed onto the surface is to treat them as a two-dimensional classical ideal gas, where the energy of an adsorbed atom is

$$\varepsilon(\vec{p}) = \frac{|\vec{p}|^2}{2m} - \varepsilon_0,$$ where $\vec{p}$ is the two-dimensional momentum. Do not concern yourself with the details of the binding. Treat ε_0 as a known parameter.

a. What is the classical partition function of the adsorbed atoms if N' atoms are bound to the surface?

b. What is the chemical potential of the adsorbed atoms?

c. What is the classical partition function of the $N - N'$ atoms in the volume of the container?

d. What is the chemical potential of the $N - N'$ atoms in the volume of the container?

e. When the atoms in the volume and those on the surface are in the equilibrium with each other, what is the average number of atoms adsorbed as a function of the temperature?

f. How many atoms are adsorbed on the walls in the limits of high and low temperatures, according to your answer to (e)? Does your answer make sense in these two limits?

Problem 5.46 Solution 5.46

Consider an idealized crystal made up of atoms that are rigid cubes of edge length a. Assume that the crystal is in the shape of a cube having edge na where $n > 2$ is an integer. Thus the total number of atoms in the crystal is $N_{tot} = n^3$.

a. Calculate N_{bulk}, N_{faces}, N_{edges}, $N_{corners}$.

b. Next, assume that these states have only nearest neighbor forces such that the bonds between neighbors have energy $-b$, where $b > 0$. An atom in the bulk will have energy $E_{bulk} = -6b/2$, since each bond is shared by two atoms. Determine the energies of the faces, edges, and corner atoms.

c. Plot the energy of the crystal as a function of N_{tot}. How big would the crystal have to be to have 99.9% of its energy accounted for by bulk atoms?

Problem 5.47 Solution 5.47

Use the method of Jacobians and Maxwell relations to find the reversible adiabatic change in temperature with applied pressure, $\left(\frac{\partial T}{\partial p}\right)_{S,N}$, in terms of measurable quantities V, T,

$$\alpha = \frac{1}{V}\left(\frac{\partial V}{\partial T}\right)_{p,N}, \quad \kappa_T = -\frac{1}{V}\left(\frac{\partial V}{\partial p}\right)_{T,N}, \quad C_V = T\left(\frac{\partial S}{\partial T}\right)_{p,N}.$$

Evaluate the result for an ideal monatomic gas.

Problem 5.48

Solution 5.48

Use the method of Jacobians to express the ratio of the adiabatic coefficient of expansion $\alpha_S = \frac{1}{V}\left(\frac{\partial V}{\partial T}\right)_{S,N}$ to the isothermal coefficient of thermal expansion $\alpha = \frac{1}{V}\left(\frac{\partial V}{\partial T}\right)_{p,N}$ in terms of

only the ratio $\gamma = \frac{C_p}{C_V}$. Evaluate the result for an ideal monatomic gas.

Problem 5.49

Solution 5.49

A single component material can exist in two phases, the alpha phase and the gamma phase. In the alpha phase, the equation of state is

$$\frac{p}{kT} = A + B\frac{\mu^\alpha}{kT}$$

and in the gamma phase it is

$$\frac{p}{kT} = C + D\left(\frac{\mu^\gamma}{kT}\right)^2$$

where A, B, C and D are positive functions of the absolute temperature T, $D > B$ and $C < A$. μ^α is the chemical potential of alpha, μ^γ is the chemical potential of gamma, p is the pressure and k is Boltzmann's constant.

a. Determine the difference in molar volume $v^\alpha - v^\gamma$ when these phases are in equilibrium.

b. Determine the pressure at which alpha and gamma will be in equilibrium.

Problem 5.50

Solution 5.50

Consider particles of spin 1 which have three quantum states with equally spaced non-degenerate energy levels separated by energy ε, with $\varepsilon_1 = 0$, $\varepsilon_2 = \varepsilon$, and $\varepsilon_3 = 2\varepsilon$. Consider a system of $N = 4$ particles and assume that the particles have negligible interaction energy and are distinguishable by their position in a solid. Make a table of the microstates of this system that correspond to a total energy of $E = 5\varepsilon$. Determine the total number of microstates $\Omega(N, E)$ for the given values of E and N. If the particles were indistinguishable, how many distinct microstates would there be?

Problem 5.51

Solution 5.51

The volume and area of a hypersphere of radius R and dimensionality n are given by:

$$V_n = \frac{\pi^{n/2}}{\left(\frac{n}{2}\right)!} R^n \quad S_n = \frac{n\pi^{n/2}}{\left(\frac{n}{2}\right)!} R^{n-1}.$$

Calculate exactly the fraction of the volume of a hypersphere that lies within ΔR of its surface. Equate this fraction to $\varepsilon - 1$ (where ε is a very small quantity) and solve the resulting equation exactly for $\Delta R / R$ in terms of ε and n.

Problem 5.52

Solution 5.52

Consider an extreme relativistic gas characterized by the single-particle energy states $\varepsilon\left(n_x, n_y, n_z\right) = \frac{hc}{2L}\sqrt{n_x^2 + n_y^2 + n_z^2}$. Show that $\gamma = \frac{C_p}{C_V} = \frac{4}{3}$ and that the ideal gas law is satisfied.

Problem 5.53

Solution 5.53

The energy levels of a quantum-mechanical, one dimensional, anharmonic oscillator may be approximated as

$$\varepsilon_n = \hbar\omega\left(n + \frac{1}{2}\right) - x\hbar\omega\left(n + \frac{1}{2}\right)^2,$$ where the parameter $x \ll 1$ represents the degree of

anharmonicity. Show that, to the first order in x and the fourth order in $u = \frac{\hbar\omega}{kT}$, the specific heat of a system of N such oscillators is given by

$$C = Nk\left[\left(1 - \frac{1}{12}u^2 + \frac{1}{240}u^4\right) + 4x\left(\frac{1}{u} + \frac{u^3}{80}\right)\right].$$

Solutions

Statistical and thermal physics

Solution 5.01

Problem 5.01

a. $S = -k \sum_{i=1}^{N} p_i \ln p_i$

The entropy is minimum when one of the p values $= 1$ and the rest are 0. Then, $S = 0$ because the system is fully defined (there is no missing information). To find the maximum, look at the partial derivative with respect to one of the p_i:

$$\sum_i p_i = 1$$

$$p_1 = 1 - \sum_{i=2} p_i$$

$$\frac{\partial S}{\partial p_1} = -k \left([1 + \ln p_1] - \sum_{i=2}^{n} [1 + \ln p_i] \right) = 0$$

$$1 + \ln p_1 = \sum_{i=2}^{n} [1 + \ln p_i]$$

Similarly, for all other p_j:

$$1 + \ln p_j = \sum_{i \neq j}^{n} [1 + \ln p_i]$$

So all p_i are equal to $\frac{1}{n}$.

Take the second derivative to ensure it is a maximum. (It is.)

b. Take $k = 1$. Then for the first case, $S \sim 1.39$, and for the second case, $S \sim 0.69$. This makes sense because the system is more fully defined in the second case; at least one of the states never occurs, and one of the others is very probable.

Solution 5.02

Problem 5.02

Look at the integral representation of the factorial function:

$$n! = \int\limits_0^\infty e^{-x} x^n dx$$

It is hard to get a good Taylor expansion of this function because it has a sharp peak. Instead, look at:

$$\ln\left(e^{-x} x^n\right) = n\ln x - x$$

Expand around $x_0 = n$ to give:

$$n\ln x - x \approx \ln n - n - \frac{1}{2}\frac{x^2}{n^2}$$

$$n! \approx \int\limits_0^\infty \exp\left(\ln n - n - \frac{1}{2}\frac{x^2}{n^2}\right) dx = \int\limits_0^\infty \exp\left(\ln n - n - \frac{1}{2}\frac{x^2}{n^2}\right) dx$$

$$= e^{-n} n^n \sqrt{2\pi n} \text{ as expected.}$$

Solution 5.03

Problem 5.03

In the Grand Canonical ensemble, we can fix an average total number of particles N in a volume V, where N and V are part of a larger system and free to exchange particles with the outside system. If the number of particles in the outside system is N', and the volume is V', we can define:

$$\frac{N'V}{V'} = N$$

Probability that N particles are in volume V: $\left(\dfrac{V}{V'}\right)^N$

Probability that all the rest of the particles are not in volume V:

$$\left(\frac{V'-V}{V'}\right)^{N'-N}$$

So total probability of having exactly N particles in volume V is the product of those two probabilities times the degeneracy factor as given by the binomial coefficient:

$$p_N = \frac{N'!}{N!(N'-N)!}\left(\frac{V}{V'}\right)^{N}\left(\frac{V'-V}{V'}\right)^{N'-N}$$

Because N' corresponds to the "rest of the universe," we can assume that:

$$N' \gg N$$

$$N' - N \approx N'$$

$$\Rightarrow p_N = \frac{1}{N!}\left(\frac{N'V}{V'}\right)^{N}\left(1 - \frac{V}{V'}\right)^{N'} = \frac{1}{N!}(N)^{N}\left(1 - \frac{N}{N'}\right)^{N'}$$

Now use:

$$\lim_{N'\to\infty}\left(1 - \frac{N}{N'}\right)^{N'} = e^{-N} \Rightarrow p_n = \frac{N^N}{N!}e^{-N},$$

which is just the Poisson distribution.

Solution 5.04

Problem 5.04

Density of states:

$$dN = \frac{gV}{(2\pi)^2}d^2k$$

where g = spin degeneracy factor

$$p = \hbar k$$

$$\varepsilon = \frac{p^2}{2m}$$

$$d^2k = kdkd\theta$$

giving

$$dN = \frac{gV}{(2\pi)^2} 2\pi \frac{2m}{\hbar^2} d\varepsilon$$

$$\rho(\varepsilon) = \frac{dN}{d\varepsilon} = \frac{mg}{2\pi\hbar^2}$$

Fermi energy:

$$\rho(\varepsilon)\varepsilon_F = N$$

$$\varepsilon_F = \frac{2\pi N\hbar^2}{mgA} = \frac{2\pi n\hbar^2}{mg}$$

Chemical potential:

$$n = \int_0^\infty d\varepsilon \rho(\varepsilon) \frac{1}{\exp\left[\beta(\varepsilon - \mu)\right] + 1} = \frac{mg}{2\pi\hbar^2} \int_0^\infty d\varepsilon \frac{1}{\exp\left[\beta(\varepsilon - \mu)\right] + 1}$$

$$\varepsilon_F = \int_0^\infty d\varepsilon \frac{1}{\exp\left[\beta(\varepsilon - \mu)\right] + 1} = \mu + k_B T \ln\left[\exp(-\beta\mu) + 1\right]$$

Solution 5.05

Problem 5.05

a. At $T = 0$, $PV = 2/3E$. Set pressures in each side the same. From the first problem, we know the density of states, and at $T = 0$:

$$E = \int_0^{\varepsilon_F} d\varepsilon\,\varepsilon\rho(\varepsilon) = \frac{gV}{2\pi^2}\frac{\sqrt{2}m^{3/2}}{\hbar^3}\frac{2}{5}\varepsilon_F^{5/2}$$

$$N = \frac{gV}{2\pi^2}\frac{\sqrt{2}m^{3/2}}{\hbar^3}\frac{2}{5}\varepsilon_F^{3/2}$$

Set pressure equal to get:

$$\frac{\varepsilon_{F1}}{\varepsilon_{F2}} = \left(\frac{2^{3/2}g_2}{g_1}\right)^{2/5}$$

Now get the number density $n = \frac{N}{V}$:

$$\frac{n_1}{n_2} = \frac{1}{2^{3/2}}2^{9/10}(g_1 g_2)^{2/5}$$

The spin factors are $2s + 1$.

$g_1 = 2$ for the spin-1/2 gas

$g_2 = 4$ for the spin-3/2 gas

Gives:

$$\frac{n_1}{n_2} = \frac{1}{2}$$

b. At $T = 0$ there is a Bose condensation, so the value is infinite. At high temperature the answer is the same as in the fermion case.

Solution 5.06

Problem 5.06

The expression for the density of states in terms of momentum remains the same as in the non-relativistic case; what is different is the energy-momentum relation, which becomes $E = cp$.

$$\frac{N}{V} = \frac{1}{\pi^2 \hbar^2} \int_0^{p_F} p^2 dp = \frac{V p_F^3}{3\pi^2 \hbar^2}$$

$$p_F = \left(3\pi^2\right)^{1/3} \left(\frac{N}{V}\right)^{1/3}$$

$$\varepsilon_F = c p_F = \left(3\pi^2\right)^{1/3} \hbar c \left(\frac{N}{V}\right)^{1/3}$$

Solution 5.07 Problem 5.07

In d dimensions, we can write:

$$\frac{N}{V} = \frac{1}{\left(2\pi\hbar\right)^d} \int d^d p \frac{1}{e^{\beta(\varepsilon - \mu)} - 1} = \frac{S_d}{\left(2\pi\hbar\right)^d} \int dp \frac{p^{d-1}}{e^{\beta(cp - \mu)} - 1}$$

where S_d is a constant related to integrating over the d-dimensional angular elements. For all values of $d > 1$, the integral converges, so yes Bose condensation can occur. To get the critical temperature, set $\mu = 0$:

$$\frac{N}{V} = \frac{S_d}{\left(2\pi\hbar\right)^d} \int dp \frac{p^{d-1}}{e^{\beta_c cp} - 1}$$

The integral can be solved in closed form for the given dimension.

Solution 5.08 Problem 5.08

$$\frac{dN}{d\varepsilon} = \frac{dN}{dk} \frac{dk}{d\varepsilon}$$

$$\frac{dk}{d\varepsilon} \propto k^{1-x} \propto \varepsilon^{\frac{1}{x} - 1}$$

$$V = \left(\frac{2\pi}{L}\right)^d$$

$$dN \propto L^d d^d k$$

So:

$$\frac{N}{V} \propto \int_{\infty}^{0} \frac{\varepsilon^{\frac{d}{x}-1}}{e^{\beta\varepsilon} - 1} d\varepsilon$$

The integral must not blow up, so must have the numerator go to zero at least as quickly as the denominator as $\varepsilon \to 0$. This only occurs when $\frac{d}{x} > 1$, or $d > x$. Note that the familiar case is $d = 3$, $x = 2$.

Solution 5.09

Problem 5.09

For any fluid with a volume element moving in the x-direction with velocity v, we can write the equation of motion:

$$F = ma$$

$$\frac{dP}{dx} + \rho\frac{\partial v}{\partial t} = 0$$

Mass conservation:

$$\frac{\partial \rho}{\partial t} = -\frac{\partial(\rho v)}{\partial x}$$

Small disturbance:

$$P \to P + \delta P$$

$$v \to \delta v$$

$$\rho \to \rho + \delta \rho$$

This allows us to write:

$$P + \delta P = f(\rho + \delta\rho) \approx f(\rho) + \left.\frac{df}{d\rho}\right|_\rho \delta\rho$$

$$\delta P = \left.\frac{dP}{d\rho}\right|_\rho \delta\rho \equiv \kappa\delta\rho$$

Now to get the wave equation:

$$\frac{d(P + \delta P)}{dx} + (\rho + \delta\rho)\frac{\partial(\delta v)}{\partial t} = 0$$

$$\frac{\partial(\rho + \delta\rho)}{\partial t} = -\frac{\partial\big([\rho + \delta\rho]\delta v\big)}{\partial x}$$

Keep only the terms linear in δ:

$$\frac{\partial(\delta\rho)}{\partial t} = -\rho\frac{\partial(\delta v)}{\partial x}$$

$$\rho\frac{\partial(\delta v)}{\partial t} = -\frac{\partial(\delta P)}{\partial x}$$

Use these equations to eliminate δv:

$$\rho\frac{\partial(\delta v)}{\partial t} = -\kappa\rho\frac{\partial(\delta\rho)}{\partial x}$$

$$\frac{\partial^2(\delta\rho)}{\partial t^2} = -\rho\frac{\partial}{\partial x}\frac{\partial(\delta v)}{\partial t} = \kappa\frac{\partial^2(\delta\rho)}{\partial x^2}$$

$$\frac{\partial^2(\delta\rho)}{\partial x^2} = -\frac{1}{\kappa}\frac{\partial^2(\delta\rho)}{\partial t^2}$$

So the speed of sound is $\dfrac{1}{\sqrt{\kappa}}$.

Solution 5.10

The equation of state for a Fermi/Bose gas is given by:

$$\Omega = -PV = -kT \sum_k \ln\left(1 \pm \exp\left[\beta(\mu - \varepsilon_k)\right] \right)$$

$$= -kT \int \ln\left(1 \pm \exp\left[\beta(\mu - \varepsilon_k)\right] \right) \rho(\varepsilon) d\varepsilon$$

where ρ is the density of states as calculated previously:

$$\rho(\varepsilon) = \frac{Vgm^{3/2}}{\sqrt{2}\pi^2\hbar^3} \varepsilon^{1/2}, \text{ and } \beta = \frac{1}{k_B T}.$$

This gives the integral:

$$\Omega = \mp \frac{Vgm^{3/2}}{\sqrt{2}\pi^2\hbar^3} \int \ln\left(1 \pm \exp\left[\beta(\mu - \varepsilon_k)\right] \right) \varepsilon^{1/2} d\varepsilon$$

$$= -\frac{2}{3} \frac{Vgm^{3/2}}{\sqrt{2}\pi^2\hbar^3} \int_0^\infty \frac{\varepsilon^{3/2}}{\exp\left[\beta(\varepsilon - \mu)\right]} d\varepsilon$$

Make the substitution $x = \beta\varepsilon$ to give the form:

$$\Omega = VT^{5/2}f\left(\frac{\mu}{T}\right), \text{ where:}$$

$$f\left(\frac{\mu}{T}\right) = -\frac{2}{3} \frac{gm^{3/2}k^{5/2}}{\sqrt{2}\pi^2\hbar^3} \int_0^\infty \frac{x^{3/2}dx}{e^{x-\beta\mu} \pm 1}$$

This is a definite integral, so it's a function of $\beta\mu$ only. This is the Fermi integral, whose solution is the polylogarithm function (PolyLog).

For an adiabatic process, we can calculate the entropy:

$$\frac{S}{N} = \frac{\left.\frac{d\Omega}{dT}\right|_{V,\mu}}{\left.\frac{d\Omega}{d\mu}\right|_{V,T}} = \text{constant} = \frac{\frac{5}{2}VT^{3/2}f + VT^{5/2}\frac{\partial f}{\partial T}}{VT^{5/2}\frac{\partial f}{\partial \mu}}$$

Taking the derivative gives:

$$\frac{\partial f}{\partial \mu} = -\frac{T}{\mu}\frac{\partial f}{\partial T}$$

So this gives:

$$\frac{\frac{5}{2}VT^{3/2}f + VT^{5/2}\frac{\partial f}{\partial T}}{VT^{5/2}\frac{\partial f}{\partial \mu}} = \text{constant}$$

$$\frac{5}{2}\frac{1}{T}\frac{f}{\frac{\partial f}{\partial \mu}} - \frac{\mu}{T} = \text{constant}$$

$$\frac{5}{2}k\frac{\text{PolyLog}\left(\frac{5}{2}, -e^{\beta\mu}\right)}{\text{PolyLog}\left(\frac{3}{2}, -e^{\beta\mu}\right)} - \frac{\mu}{T} = \text{constant}$$

This can only be true if $\beta\mu = $ constant. In this case:

$$\frac{N}{V} \propto \frac{1}{V}(VT)^{3/2}$$

$$\frac{N}{(VT)^{3/2}} = \text{constant}$$

$$(VT)^{3/2} = \text{constant}$$

$$\Omega = -PV \propto (VT)^{5/2}$$

$$PV\left(V^{2/3}\right) = \text{constant}$$

$$PV^{5/3} = \text{constant}$$

Solution 5.11

Choosing the electric field to point in the z-direction, we can write the Hamiltonian:

$$H = \frac{\vec{p}^2}{2m} - \vec{\mu} \cdot \vec{E} = \frac{\vec{p}^2}{2m} - \mu E \cos\theta$$

$$P = \frac{\exp\left(-\beta \sum_i \left[\frac{p_i^2}{2m} - \mu E \cos\theta_i\right]\right)}{\exp\left(-\beta \sum_i \left[\frac{p_i^2}{2m} - \mu E \cos\theta_i\right]\right) \prod_i d^3 r_i d^3 p_i d(\cos\theta_i) d\phi_i}$$

The only part that contributes to the polarization density is the dipole energy.

$$\mu \cos\theta_i = \frac{\int\limits_{-1}^{1} d(\cos\theta_i) \mu \cos\theta_i e^{\beta\mu E \cos\theta_i}}{\int\limits_{-1}^{1} d(\cos\theta_i) e^{\beta\mu E \cos\theta_i}} = \mu \left[\coth(\beta\mu E) - \frac{1}{\beta\mu E}\right]$$

Giving:

$$P = \frac{N\mu}{V}\left[\coth(\beta\mu E) - \frac{1}{\beta\mu E}\right]$$

Solution 5.12

Calculate the canonical partition function for a gas of N identical particles:

$$Z = \frac{q^N}{N!}$$

$$q = \left(\frac{2\pi m}{\beta h^2}\right)^{3/2} V$$

Now assume a hard-core potential:

$$u(r) = \begin{cases} \dfrac{-\left(\dfrac{d}{r}\right)^6 \text{ for } r \geq d}{\infty \text{ for } r < d} \end{cases}$$

The value of q for each particle must now be changed to accommodate (a) the inaccessible volume in the hard spheres and (b) the mean-field interaction. We can write:

$$q = \left(\frac{2\pi m}{\beta h^2}\right)^{3/2}\left(V - \frac{2}{3}\pi d^3\right)e^{-\beta\varphi/2} \equiv \left(\frac{2\pi m}{\beta h^2}\right)^{3/2}(V - nb')e^{-\beta\varphi/2}$$

where

$$\varphi = \int_d^\infty u(r)\frac{N}{V}4\pi r^2 dr = -2\frac{N}{V}\varepsilon b' \equiv -2\frac{N}{V}a'$$

$$\ln Z = N\ln\left(V - Nb'\right) + \frac{\beta N^2 a'}{V} - N\ln\left(\frac{\beta h^2}{2\pi m}\right)^{3/2} - \ln N!$$

To get the van der Waals equation, take the derivative:

$$P = \frac{1}{\beta}\frac{\partial \ln Z}{\partial V} = \frac{N}{\beta(V - Nb')} - \frac{N^2 a'}{V^2}$$

Solve for kT:

$$\left(P + \frac{N^2 a'}{V^2}\right)(V - Nb') = NkT$$

Solution 5.13

Problem 5.13

$$\frac{C_V}{T} = \left(\frac{\partial S}{\partial T}\right)_V, \quad \left(\frac{\partial S}{\partial V}\right)_T = \left(\frac{\partial P}{\partial T}\right)_V$$

$$\Rightarrow \left(\frac{\partial C_V}{\partial V}\right)_T = T\left(\frac{\partial^2 P}{\partial T^2}\right)_V = 0$$

Solution 5.14

Problem 5.14

Expand P about n_c and T_c in a 2-dimensional Taylor series:

$$P = \frac{nkT}{1-nb} - an^2$$

$$P(n,T) \approx P(n_c,T_c) + \left.\frac{\partial P}{\partial n}\right|_{n_c,T_c}(n-n_c) + \left.\frac{\partial P}{\partial T}\right|_{n_c,T_c}(T-T_c) + \left.\frac{\partial^2 P}{\partial n \partial T}\right|_{n_c,T_c}(n-n_c)(T-T_c)$$

$$+\frac{1}{2}\left.\frac{\partial^2 P}{\partial n^2}\right|_{n_c,T_c}(n-n_c)^2 + \frac{1}{6}\left.\frac{\partial^3 P}{\partial n^3}\right|_{n_c,T_c}(n-n_c)^3$$

Note that the first and second derivatives with respect to n vanish at the critical point. There are no higher order terms in T because the equation is linear in T. Thus, to 3rd order in n, this reduces to:

$$P(n,T) \approx P(n_c,T_c) + \left.\frac{\partial P}{\partial T}\right|_{n_c,T_c}(T-T_c) + \left.\frac{\partial^2 P}{\partial n \partial T}\right|_{n_c,T_c}(n-n_c)(T-T_c) + \frac{1}{6}\left.\frac{\partial^3 P}{\partial n^3}\right|_{n_c,T_c}(n-n_c)^3$$

Take the derivatives and plug in:

$$\left.\frac{\partial P}{\partial T}\right|_{n_c,T_c} = \frac{n_c k}{1-n_c k} \equiv \alpha$$

$$\left.\frac{\partial^2 P}{\partial n \partial T}\right|_{n_c,T_c} = \frac{k}{\left(1-n_c b\right)^2} \equiv \beta$$

$$\frac{1}{6}\left.\frac{\partial^3 P}{\partial n^3}\right|_{n_c,T_c} = \frac{kT_c b^2}{\left(1-n_c b\right)^4} \equiv \gamma$$

169

$$\Rightarrow P = \alpha(T - T_c) + \beta(T - T_c)(n - n_c) + \gamma(n - n_c)^3$$

Get the isothermal compressibility:

$$\kappa = -\frac{1}{V}\frac{\partial V}{\partial P}\bigg)_T = -\frac{1}{V}\frac{1}{\frac{\partial V}{\partial P}\big)_T}$$

$$\frac{\partial V}{\partial P}\bigg)_T = -\beta\frac{(T - T_c)N}{V^2} + 3\gamma(n - n_c)^2\left(-\frac{N}{V^2}\right)$$

$$\kappa = \frac{1}{\beta(T - T_c)N + 3\gamma N(n - n_c)^2} \xrightarrow{n \to n_c} \frac{1}{|T - T_c|}$$

To see the dependence of $n - n_c$ on $T - T_c$, look at:

$$\frac{\partial P}{\partial V}\bigg)_T = 0 = -\beta\frac{(T - T_c)N}{V^2} + 3\gamma(n - n_c)^2\left(-\frac{N}{V^2}\right)$$

$$\beta(T - T_c) = 3\gamma(n - n_c)^2$$

$$(n - n_c)^2 \propto \sqrt{(T - T_c)}$$

Solution 5.15

Problem 5.15

$$-PV = \Omega = \mp kT\sum_i \ln\left(1 \pm e^{-\beta(\varepsilon_i - \mu)}\right)$$

In the classical limit:

$$e^{\beta\varepsilon_i + \alpha} \gg 1$$

$$n_i \ll 1$$

$$n_i = \frac{1}{e^{\beta \varepsilon_i + \alpha} + 1} \approx e^{-\beta \varepsilon_i - \alpha} \text{ for both fermions and bosons}$$

Use the identity:

$\ln(1 + x) \approx x$ for $x \ll 1$ to give

$$\Omega \approx -kT \sum_i e^{-\beta \varepsilon_i - \alpha} = -NkT$$

$$PV = NkT$$

Solution 5.16

Problem 5.16

$T_\lambda = 2.15$ K, $p_\lambda = 0.146$ g/cm³

Compare with T_c:

$$\frac{dN}{dE} = \frac{dN}{dk}\frac{dk}{dE} = \left(\frac{V}{2\pi^2}k^2\right)\frac{dk}{dE} = \frac{V}{2\pi^2}\frac{m}{\hbar^2}d\varepsilon\sqrt{\frac{2m}{\hbar^2}\varepsilon}$$

$$N = \int B(\varepsilon)D(\varepsilon)$$

where B = Bose distribution; D = density of states as above. This gives:

$$N = \frac{V}{4\pi^2}\left(\frac{2m}{\hbar^2}\right)^{3/2}\int_0^\infty \frac{\varepsilon^{1/2}d\varepsilon}{e^{(\varepsilon-\mu)\beta} - 1}$$

Take $\mu \to 0$:

$$\frac{N_c}{V} = \frac{1}{4\pi^2}\left(\frac{2mkT_c}{\hbar^2}\right)^{3/2}\Gamma\left(\frac{3}{2}\right)\zeta\left(\frac{3}{2}\right) \approx 2.61\left(\frac{mkT_c}{2\pi\hbar^2}\right)^{3/2}$$

$$T_c \approx \frac{h^2}{2\pi mk}\left(\frac{N}{2.61V}\right)^{2/3} \approx 3.2K$$

The temperature for the superfluid phase transition is a little lower than the Bose condensation temperature at this density. Whatever similarity there is between the two is almost entirely accidental: Bose condensation and lambda transition are very dissimilar. Bose condensation is a first order phase transition whose critical temperature goes up with density; the opposite is true for the lambda transition, which is a critical transition running from the melting curve to the vapor pressure curve. Superfluidity requires Bose condensation, but Bose condensation does not necessarily imply superfluidity.

Solution 5.17 Problem 5.17

Repeat the calculation for the Bose gas, but with the Fermi distribution:

$$N = \frac{V}{4\pi^2}\left(\frac{2m}{\hbar^2}\right)^{3/2}\int_0^\infty \frac{\varepsilon^{1/2}d\varepsilon}{e^{(\varepsilon-\mu)\beta} + 1}$$

Take $\mu \to 0$:

$$\frac{N_c}{V} = \frac{1}{4\pi^2}\left(\frac{2m}{\hbar^2}\right)^{3/2}\frac{\pi^2 k^2 T_c^2}{24} = \frac{k^2 T_c^2}{96}\left(\frac{2m}{\hbar^2}\right)^{3/2}$$

$$T_c^2 \approx \frac{96}{k^2}\frac{N_c}{V}\left(\frac{\hbar^2}{2m}\right)^{3/2}$$

Where $\mu = 0$ for a Fermi gas, we get:

$$n(\varepsilon) = \frac{1}{e^{\varepsilon\beta} + 1}$$

$$N = D\int_0^\infty \frac{\varepsilon^{1/2}d\varepsilon}{e^{\varepsilon\beta} + 1}$$

$$PV = \frac{2}{3}\int d\varepsilon n(\varepsilon)D(\varepsilon)$$

All of these expressions are finite without singularities, as are their derivatives. Thus there are no phase transitions or unusual behavior of any thermodynamic quantities at this value.

Solution 5.18

Problem 5.18

An ergodic system is one where all points in phase space are eventually visited.

For 2 uncoupled harmonic oscillators, the allowed phase space is a torus:

$$E = E_1 + E_2$$

$$p_1 = \sqrt{2mE}\cos\varphi_1$$

$$q_1 = \sqrt{\frac{2E}{k_1}}\sin\varphi_1$$

$$p_2 = \sqrt{2mE}\cos\varphi_2$$

$$q_2 = \sqrt{\frac{2E}{k_2}}\sin\varphi_2$$

φ_1 has a period $\frac{2\pi}{\omega_1}$ so it comes to a complete circuit at $t = \frac{2\pi n}{\omega_1}$, where n is an integer. At the same time, φ_2 is behaving like:

$$\varphi_2 = \varphi_{02} + \omega_2\left(\frac{2\pi n}{\omega_1}\right)$$

If ω_2/ω_1 is an integer, the path will be a closed helix in phase space. But if this ratio is irrational, the oscillators never meet and never visit the same point twice, filling all phase space as $t \to \infty$.

Solution 5.19

Problem 5.19

The quantum Hall effect refers to the quantization of the current density magnitude of a 2D electron gas subject to a strong magnetic field. That is, the current density j, which is perpendicular to E, has the form:

$$j_i = \sum_j \sigma_{ij} E_j$$

where σ is the conductivity tensor

$$\sigma = \begin{pmatrix} 0 & \frac{-ne^2}{h} \\ \frac{ne^2}{h} & 0 \end{pmatrix}$$

n = integer

So the magnitude of j/E is given by σ_{xy} and the diagonal terms vanish (there is no dissipation). The system is this related to a superfluid or superconducting system. For ideal 2D electrons, the B field makes each electron have discrete energy levels equal to the classical cyclotron frequency times an integer times Planck's constant; these are called "Landau levels," and at sufficiently low T, are all either completely full or completely empty. The filled ones contribute a current density perpendicular to E of $\frac{e^2 E}{h}$. For low T:

$$\frac{\text{Hall current}}{V} = \frac{ne^2}{h}$$

where n is an integer. These integer steps have been experimentally verified.

Solution 5.20

Problem 5.20

a. Calculate the pressure as:

$$P = -\left(\frac{\partial \Omega}{\partial V}\right)_{T,\mu} = \frac{1}{3\pi^2}\left(\frac{2m}{\hbar^2}\right)^{3/2} \int_0^\infty \frac{d\varepsilon\, \varepsilon^{3/2}}{e^{\beta(\varepsilon - \mu)} + 1}$$

$$P \approx \frac{1}{3\pi^2} \left(\frac{2m}{\hbar^2} \right)^{3/2} \left[\int_0^\infty \frac{d\varepsilon\,\varepsilon^{3/2}}{e^{\beta(\varepsilon-\mu)}+1} + \frac{\beta(\varepsilon-\varepsilon_F)}{4} \int_0^\infty d\varepsilon\,\varepsilon^{3/2} \text{sech}^2 \left\{ \frac{\beta}{2}(\varepsilon-\varepsilon_F) \right\} \right]$$

Use the Sommerfeld expansion to give approximate values for the integrals:

$$P \approx \frac{1}{3\pi^2} \left(\frac{2m}{\hbar^2} \right)^{3/2} \left[\frac{2}{5}\varepsilon_F^{5/2} + \frac{\pi^2}{4\beta^2}\varepsilon_F^{1/2} + \frac{\beta}{4}(\varepsilon-\varepsilon_F)\frac{4}{\beta}\varepsilon_F^{3/2} \right] = \frac{2}{15\pi^2} \left(\frac{2m}{\hbar^2} \right)^{3/2} \varepsilon_F^{5/2} \left[1 + \frac{5\pi^2}{12\beta^2\varepsilon_F^2} \right]$$

Substitute:

$$\frac{N}{V} = \frac{1}{3\pi^2} \left(\frac{2m}{\hbar^2} \right)^{3/2} \varepsilon_F^{3/2} \text{ to give:}$$

$$P = \left(\frac{N}{V} \right)^{5/3} \left(\frac{\hbar^2}{m} \right) \frac{(3\pi^2)^2}{5}, \text{ where } m \text{ is the electron mass. Note that this is independent of}$$

temperature.

b. In the non-degenerate case, use the classical equation of state:

$$-\Omega = \frac{V}{3\pi^2} \left(\frac{2m_{He}}{\hbar^2} \right)^{3/2} \int_0^\infty d\varepsilon\,\varepsilon^{3/2} e^{\beta(\varepsilon-\mu+W)} = \frac{V}{\sqrt{2\pi^3}\beta^{5/2}} \left(\frac{m_{He}}{\hbar^2} \right)^{3/2} e^{\beta(\mu-W)}$$

$$\Rightarrow P = \frac{V}{\sqrt{2\pi^3}\beta^{5/2}} \left(\frac{m_{He}}{\hbar^2} \right)^{3/2} e^{\beta(\mu-W)} - \frac{1}{5\sqrt{\pi}} \left(\frac{m_{He}}{\beta\hbar^2} \right)^{3/2} \left(\frac{\sqrt{2}}{3\pi^2} \right)^{1/3} e^{\beta(\mu-W)}$$

Substitute:

$$\frac{N_{He}}{V} = \frac{1}{\sqrt{2}} \left(\frac{m_{He}}{\pi\beta\hbar^2} \right)^{3/2} e^{\beta(\mu-W)}$$

To give:

$$P = \frac{N_{He}}{\beta V} - \frac{1}{5} \left(\frac{4\pi}{3} \right)^{1/3} \frac{GM^2}{V^{4/3}}$$

The first term is the deal classical gas term and can be neglected.

c. Set the values of P in parts (a) and (b) equal. Use the fact that there are two electrons per He atom, and the mass of each atom is approximately that of the nuclei, to make the substitutions:

$$N = 2N_{He} = \frac{M}{2m_{He}}$$

To give the final result:

$$RM^{1/3} = \frac{3^{4/3}\pi^{2/3}\hbar^2}{8Gmm_{He}^{5/3}}$$

Solution 5.21

Problem 5.21

Hold the protons fixed, displace all of the electrons by an amount x. This creates a negative charge on one side and positive charge on the other that is proportional to x and the plasma density n: $q = \pm enx$. The acceleration caused by an E field (neglecting proton motion) is:

$$\frac{d^2x}{dt^2} = \frac{Ee}{m_e} = -\frac{e^2n}{\varepsilon_0 m_e}x$$

This is just a harmonic oscillator equation with frequency $\omega = \sqrt{\dfrac{e^2n}{\varepsilon_0 m_e}}$.

Solution 5.22

Problem 5.22

The formula for the resistivity in terms of mean free path is:

$$\rho = \frac{v_t h m_e}{n_e^2 \lambda}, \text{ where } n_e \text{ is the electron density and } v_t \text{ is the velocity. Classically:}$$

$$v_t h = \sqrt{\frac{3kT}{m_e}} \Rightarrow \rho = \frac{\sqrt{3m_e kT}}{n_e^2 \lambda} \propto T^{3/2}$$

Quantum mechanically, the velocity is the Fermi velocity which is T-independent. Thus the dependence of the resistivity on temperature is the same as that of the reciprocal of the mean free path, i.e. T.

Solution 5.23

Problem 5.23

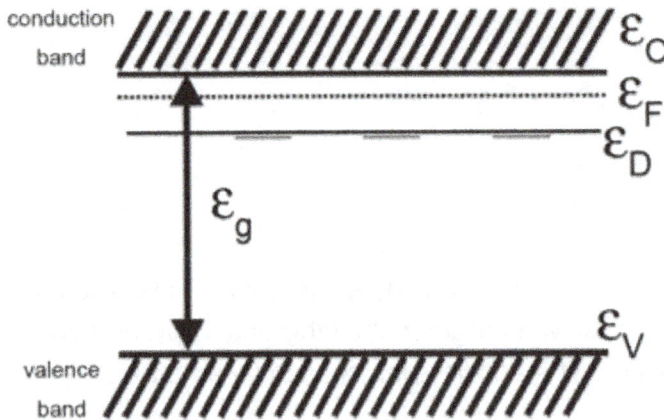

$$N_d = \int_0^\infty \frac{8\pi\sqrt{2m}}{h^3} \frac{d\varepsilon\sqrt{\varepsilon}}{e^{\beta(\varepsilon + \varepsilon_C - \varepsilon_D - \varepsilon_F)} + 1}$$

If we can assume that $\varepsilon_C - \varepsilon_F \gg kT$, then:

$$N_d \approx \frac{2(2\pi m k T)^{3/2}}{h^3} e^{-\beta(\varepsilon_C - \varepsilon_D - \varepsilon_F)}$$

Solution 5.24

Problem 5.24

a. $Z_{vib} = \dfrac{1}{1 - e^{-\beta\varepsilon_{vib}}} = \sum_{\infty}^{n=0} e^{-\beta n \varepsilon_{vib}}$

$$E_{vib} = -\frac{\partial}{\partial \beta} \ln Z_{vib} = \frac{e^{-\beta \varepsilon_{vib}}}{e^{-\beta \varepsilon_{vib}} - 1}$$

b. For non-identical particles, the first two terms will represent $\ell = 0$ and $\ell = 1$. For identical bosons, only even values of ℓ will contribute. So:

    i. $1 + 3e^{-2\beta \varepsilon_{rot}}$

    ii. $1 + 5e^{-6\beta \varepsilon_{rot}}$

c. There are 2 degrees of freedom for rotations: the x and y axes. So according to the equipartition theorem, the average energy is kT.

Solution 5.25   Problem 5.25

The surface gravitational acceleration is about $g = 1.67$ m²/s. Escape velocity is given by $(2gr)^{1/2}$. The RMS thermal velocity is $(2kT/m)^{1/2}$. Equate the two to get the limiting mass of $m > 2.14 \times 10^{-27}$ kg. Given that 1 atom mass unit is 1.66×10^{-27} kg, this is a very low limit, lighter even that H_2. To realize why the moon has no atmosphere, we must realize that the upper tail of the Boltzmann distribution of velocities will allow heavier gases to reach escape velocity, so that over long periods of time, all of the gas can escape. There are also non-thermal escape mechanisms, such as solar wind stripping.

Solution 5.26   Problem 5.26

For each type of particle, the magnetization M is given by:

$$M = g\mu\langle S \rangle = \frac{1}{\beta}\frac{\partial \ln Z}{\partial \beta} = N_\uparrow(-\mu) + N_\downarrow(\mu) = N\mu\frac{e^{\beta\mu H} - e^{-\beta\mu H}}{e^{-\beta\mu H} + e^{\beta\mu H}}$$

At high temperatures $\beta\mu H \ll 1$, so $M \to N\beta\mu^2 H$.

$$\langle S \rangle = \hbar N \mu H \beta$$

So for the two component gas:

$$\langle S \rangle = \frac{\hbar N_a \mu_a H \beta}{2} + \frac{\hbar N_b \mu_b H \beta}{2}$$

At low temperatures, $\beta \mu H \gg 1$ and $M \rightarrow \dfrac{3}{\epsilon_F} \dfrac{N}{V} \mu^2 H$.

Substitute the expression for ϵ_F:

$$\epsilon_F = \frac{\hbar^2}{2m} \left(\frac{3\pi^2 N}{V} \right)^{2/3}$$

So for the two component gas:

$$\langle S \rangle = \frac{3^{1/3}}{\pi^{1/3} h} H V^{2/3} \left[N_a^{1/3} \mu_a m_a + N_b^{1/3} \mu_b m_b \right]$$

The condition for parallel at low/antiparallel at high is thus:

$$N_a m_a + N_b m_b < 0$$

$$N_a^{1/3} \mu_a m_a + N_b^{1/3} \mu_b m_b > 0$$

Solution 5.27

Problem 5.27

a. The number of possible arrangements is:

$$W = \frac{2N!}{\left(\frac{N}{2} + m \right)! \left(\frac{N}{2} - m \right)!}$$

b. Using Stirling's approximation, we get:

$$S = k \ln W \approx k \left[\ln 2 + N \ln N - N - \left(\frac{N}{2} + m \right) \ln \left(\frac{N}{2} + m \right) - \left(\frac{N}{2} - m \right) \ln \left(\frac{N}{2} - m \right) + N \right]$$

Now we want to expand around $l = 2mx$ to first order, assuming l is small:

$$S \approx k \left[\ln 2 + N \ln N - \left(\frac{N}{2} + \frac{l}{2x} \right) \ln \left(\frac{N}{2} + \frac{l}{2x} \right) - \left(\frac{N}{2} - \frac{l}{2x} \right) \ln \left(\frac{N}{2} - \frac{l}{2x} \right) \right]$$

$$\approx k \left[C - \frac{l^2}{Nx^2} \right], \text{ where } C \text{ is a constant.}$$

c. $dU = TdS + Fdl$ where F is the retractive force.

d. For an ideal polymer, $dU/dl = 0$, so:

$$F = -T \left(\frac{\partial S}{\partial l} \right)_{T,p} = \frac{2lT}{Nx^2}$$

Solution 5.28

Problem 5.28

Number of possible states and entropy:

$$W = 2^n \frac{N!}{(N-n)! \, n!}$$

where $n = \dfrac{E}{\varepsilon}$

$$S = k \ln W \approx k \left(n \ln 2 + N \ln N - (N-n) \ln(N-n) - n \ln n \right)$$

The temperature is given by:

$$\frac{1}{T} = \frac{\partial S}{\partial E} = \frac{1}{\varepsilon} \left(2 \left(\frac{N\varepsilon}{E} - 1 \right) \right)$$

$T < 0$ when $E > \dfrac{2}{3} N\varepsilon$.

Negative temperatures mean that the higher energy levels have higher occupation than the lower levels. This is essential for the operation of lasers.

Solution 5.29 Problem 5.29

Consider a Carnot cycle between temperatures T and $T + \Delta T$. We can assume that negligible work is done in the adiabatic expansion and compression between the isotherms. Suppose heat L is absorbed at $T + \Delta T$ to vaporize one mole of liquid; the work done is $W = P(T + \Delta T)\Delta V$. During isothermal compression at temperature T, the work done $= PT\Delta V$. So the total work done in the cycle is:

$$\left[P(T + \Delta T) - PT \right]\Delta V = L\frac{\Delta T}{T}$$

Convert to differentials:

$$\frac{dP}{dT} = \frac{L}{T\Delta V}$$

Solution 5.30 Problem 5.30

Burning is oxidation, or removal of electrons. In the alternate universe, the energy of the H-atom would double, and so would the burning temperature.

Solution 5.31 Problem 5.31

Make use of the Clausius-Clapeyron equation, assuming that the density of the gas is much less than that of the liquid.

$$\frac{dP}{dT} = \frac{L}{T\Delta V}$$

$$P\Delta V = nRT$$

$$\Rightarrow \frac{dP}{dT} = \frac{L}{T}\left(\frac{P}{nRT}\right)$$

$$nR\frac{dP}{P} = \frac{L}{T^2}dT$$

Integrate:

$$\frac{nR}{L}\ln\frac{P_f}{P_0} = \frac{1}{T_0} - \frac{1}{T_f}$$

Solution 5.32

Problem 5.32

The rms velocity is given by integral over the velocity distribution, $f(v)$:

$$\overline{v^2} = \frac{\int\limits_0^\infty f(v)v^2 dv}{\int\limits_0^\infty f(v)dv}$$

The velocity distribution of the particles that manage to escape is related to the flux at the hole, given by:

# of molecules hitting area $= \frac{1}{2}nvf(v)dv\sin\theta\cos\theta d\theta$

So the velocity distribution of effusing particles, f_e, will have an extra factor of v, skewing it to higher velocities:

$$f_e(v) \propto vf(v)$$

where $f(v)$ is the ordinary Maxwell distribution, $f(v) \propto v^2 e^{-av^2}$, where $a = \frac{m}{2kT}$.

This gives:

$$\overline{v_e^2} = \frac{\displaystyle\int_0^\infty e^{-av^2}v^5 dv}{\displaystyle\int_0^\infty e^{-av^2}v^3 dv} = \frac{2}{a}$$

Substituting gives:

$\sqrt{\overline{v_e^2}} = \sqrt{\dfrac{4kT}{m}}$, which is faster than the rms velocity of all the particles

Solution 5.33

Problem 5.33

$$e = \frac{W}{Q} = \frac{T_2 - T_1}{T_2}$$

$$W = Q\left(\frac{T_2 - T_1}{T_2}\right)$$

Solution 5.34

Problem 5.34

For specific heat:

$$dQ = \left(C_M + \frac{aH^2}{T^2}\right)dT - \left(\frac{aH}{T}\right)dH$$

$$C_H = \left.\frac{dQ}{dT}\right|_H = C_M + \frac{MH}{T}$$

For the entropy:

$$dQ = TdS$$

$$\Delta S = \int_{H_1}^{H_2} \frac{dQ}{T} = -\frac{a}{T^2}\int_{H_1}^{H_2} H dH = \frac{a}{2T^2}\left(H_1^2 - H_2^2\right)$$

Solution 5.35

Problem 5.35

$U = Vu$, u = energy density

$$P = \frac{1}{3}u$$

$$TdS = dU + pdV = udV + Vdu + pdV = \frac{4}{3}UdV + V\left(\frac{du}{dT}\right)dT$$

$$\frac{\delta S}{\delta V} = \frac{4u}{3T}$$

$$\frac{\delta S}{\delta T} = \frac{V}{T}\frac{du}{dT}$$

$$\frac{\partial^2 S}{\partial V \partial T} = \frac{4}{3}\left(\frac{1}{T}\frac{du}{dT} - \frac{u}{T^2}\right) = \frac{1}{T}\frac{du}{dT} \Rightarrow \frac{du}{dT} = 4\frac{u}{T}$$

$$\frac{du}{u} = 4\frac{dT}{T} \text{ so } \ln u = C + \ln T^4 \text{ or } u = cT^4 \text{ where } c \text{ is constant.}$$

If the cavity expands adiabatically:

$$TdS = 0 = \frac{4}{3}udV + Vdu = \frac{4}{3}cT^4dV + 4cVT^3dT$$

$$\frac{dT}{T} = -\frac{1}{3}\frac{dV}{V}$$

$$T \propto V^{-1/3}$$

So doubling the length of a side halves the temperature. The energy density u is reduced by a factor of 16, V increases by 8, so the total internal energy decreases by a factor of 2, which is the work done.

Solution 5.36

Problem 5.36

The entropy is simply given by the number of ways to make a given arrangement, namely:

$$S = k\left[\log\frac{N!}{(N-n)!n!} + \frac{M!}{(M-m)!n!}\right]$$

Using Stirling's approximation:

$$S \approx k\left[N\log N + M\log M - (N-n)\log(N-n) - (M-n)\log(M-n) - 2n\log n\right]$$

The energy to move an atom can be calculated from the equilibrium condition:

$$\frac{\partial U}{\partial n} = T\frac{\partial S}{\partial n}$$

$$\frac{\partial S}{\partial n} = \log\left(\frac{N-n}{n}\right) + \log\left(\frac{M-n}{n}\right)$$

$$\Rightarrow E_1 = \left[\log\left(\frac{N-n}{n}\right) + \log\left(\frac{M-n}{n}\right)\right]$$

For $n \ll N, M$ can express:

$$E_I = kT\left[\log\left(\frac{NM}{n^2}\right)\right]$$

Solution 5.37

Problem 5.37

If the gas in the tank undergoes adiabatic expansion, with V increasing by a factor of 10, there will be no entropy change. The final temperature will be given by:

$$TV^{\gamma-1} = TV^{2/3} = \text{constant} \Rightarrow T_f = \frac{T_i}{10^{2/3}}$$

The entropy change of the gas leaving is $\Delta S = nR\ln\left(\frac{V_2}{V_1}\right)$, where n = # of moles leaving

$= 0.9(PV/RT)$. The ratio of the volumes is 100:1 since the gas in the balloon will come to equilibrium with the outside atmosphere at 1 atm. Thus:

$$\Delta S = \frac{PV}{T}\ln(100)$$

$$T\Delta S = PV \ln(100)$$

The heat added to the gas as it leaves is:

$$P\Delta V - \Delta U = P\Delta V - \frac{3}{2} R\Delta(nT)$$

$$\Delta V = 0.9(100V - V)$$

$$\Delta(nT) = n_0 T_i - 0.1 n_0 T_f$$

with temperatures calculated as above. This will be smaller than ΔS because the process is irreversible.

Solution 5.38

Problem 5.38

If all of the gravitational energy is available, we can use:

$$U = \frac{2}{5} \frac{GM^2}{R}$$ as the total energy. The lifetime is then just U/L. Plugging in the numbers gives 3.8×10^{14} s for the lifetime with gravitational contraction.

For nuclear fusion, the energy yield is 7 MeV per fusion event, or 1.12×10^{-12} J. To get the observed luminosity, we would need $4 \times 10^{26}/1.12 \times 10^{-12} = 3.57 \times 10^{38}$ events per second. Each hydrogen weighs 1.67×10^{-27} kg, so the mass consumed per second is $\sim 6 \times 10^{11}$ kg. The entire mass would be consumed in $(2 \times 10^{30}/6 \times 10^{11}) = 3.3 \times 10^{18}$ s.

To make the timescales match, the value of U would have to increase by about 10^4. If M is unchanged, then R would have to shrink by a factor of 10,000, reducing it to about 70 km! A very small sun indeed.

Solution 5.39

Problem 5.39

Model the photon gas as a harmonic oscillator. The partition function is given by:

$$Z = \sum_{n=0}^{\infty} e^{-\beta \hbar \omega \left(n + \frac{1}{2}\right)} = \frac{e^{-\beta \hbar \omega/2}}{1 - e^{-\beta \hbar \omega/2}}$$

Then the energy as a function of frequency is given by:

$$U(\omega) = -\frac{\partial \ln Z}{\partial \beta} = \frac{\hbar \omega}{2} + \frac{\hbar \omega}{\beta^{\beta \hbar \omega} - 1}$$

And the total energy:

$$U = \int_{0}^{\infty} d\omega\, U(\omega) g(\omega)$$

where g is the density of states, $g(\omega) = \dfrac{V \omega^2}{\pi^2 c^3}$.

The first term in U gives a divergent integral, but can be defined to be 0. (Do not ask why that works!) The second term is readily integrated to yield:

$$\frac{U}{V} = \frac{k^4 T^4}{\pi^4 c^3 \hbar^2} \Gamma(4) \zeta(4) = \frac{\pi^4 k^3 T^4}{15 c^3 \hbar^2}$$

This is starting to have the right form, but there's one more step to get the power emitted, which is energy times flux. Recall from Problem 32 that the flux of particles hitting a container is $\frac{1}{4} n v_{avg} = \frac{1}{4} nc$ for photons. So the power is related to the energy density by:

$$P = \hbar \omega (\text{flux}) = \hbar \omega \frac{nc}{4} = \frac{U}{V} \frac{c}{4}$$

$$P = \frac{\pi^2 k^3 T^4}{60 c^2 \hbar^2} \equiv \sigma T^4$$

Solution 5.40

Problem 5.40

First calculate the total solar flux F, at the distance of Neptune. This will be the sun's black-body energy divided over the area of the sphere with a radius at the distance of Neptune:

$$F = \frac{4\pi R_s^2 \sigma T^4}{4\pi d^2} = 1.55 \, \frac{W}{m^2}$$

(Compare to the value for the Earth of 1370 W/m². Brrr!)

Now assuming that Neptune radiates as a black body, its temperature will be given by:

$$\sigma T_n^4 = \frac{1-A}{4} F$$

where A is the albedo, and the factor of 4 arises due to shadowing. This gives us an approximate temperature of 45 K for the planet (extra brr)!

Now we need to know which elements will have thermal velocities much lower than that of the escape velocity at that temperature. We can pick a thermal velocity of 0.1 × (escape velocity) as a safe value at which the molecules will be retained in the atmosphere over the planet's history (see Problem 25). Thus we're looking for a value of m for which:

$$\sqrt{\frac{3kT}{m}} = 0.1 \sqrt{\frac{2MG}{R_n}}$$

Plugging in the values gives a value < 1 amu. So everything should be retained in Neptune's atmosphere, even hydrogen. In fact, its actual atmosphere consists of 99% H and He, although its bright blue color is caused by methane.

Solution 5.41   Problem 5.41

Yes, improvement can result because the waste heat of the engine and pump are released into the house as well.

Solution 5.42   Problem 5.42

The surface area of the box is 0.96 m² for all 6 sides, and its volume is 0.064 m³. Assume a nitrogen molecule is about 3 Å long, for an effective area of about 7×10^{-20} m² (envisioning a spinning barbell). Then the number of molecules that can fit on the surface is given by the ratio of the surface area to the molecular area, or about 1.4×10^{19} molecules. Take this value of N and plug $PV = NkT$ to solve for $P = 0.007$ torr.

Solution 5.43

a. $E_n = \hbar\omega\left(n + \dfrac{1}{2}\right)$. Ignore zero point.

$$E_n = \hbar\omega n$$

$$Z = \sum_n e^{-\beta E_n} = \sum_n e^{-\beta\hbar\omega n} = \sum_n \left(e^{-\beta\hbar\omega}\right)^n = \frac{1}{1 - e^{-\beta\hbar\omega}}$$

$$\langle E \rangle = -\frac{\partial}{\partial\beta}\ln Z = \frac{\hbar\omega e^{-\beta\hbar\omega}}{1 - e^{-\beta\hbar\omega}}$$

$$\langle E \rangle = \hbar\omega\langle N \rangle$$

$$\langle N \rangle = \frac{1}{e^{\beta\hbar\omega} - 1}$$

b. In terms of wavelength and frequency, $\omega = ck$. Then, converting the volume-wise density into an integral only over the surface:

$$\rightarrow 2\frac{V}{(2\pi)^3}\left(4\pi k^2\right)dk, \quad k = \frac{\omega}{c} \quad dk = \frac{1}{c}d\omega$$

For:

$$\rightarrow 2\frac{V}{(2\pi)^3}\left(4\pi\frac{\omega^2}{c^2}\right)\frac{1}{c}d\omega = \frac{V\omega^2}{\pi^2 c^3}d\omega.$$

c. Using the equipartition of energy, classically

$$E = \int_0^\infty kT\left(\frac{V\omega^2}{\pi^2 c^3}\right)d\omega.$$ This integral clearly diverges as ω becomes very large, and scales linearly in T.

d. $\langle E \rangle = \dfrac{\sum\limits_i E_i e^{-\beta E_i}}{Z} = \dfrac{-\sum\limits_i \frac{\partial}{\partial \beta} e^{-\beta E_i}}{Z} = -\dfrac{1}{Z}\dfrac{\partial}{\partial \beta} Z$

$$Z = \dfrac{1}{1 - e^{-\hbar\omega\beta}} \qquad \dfrac{\partial Z}{\partial \beta} = -\dfrac{\hbar\omega e^{-\hbar\omega\beta}}{\left(1 - e^{-\hbar\omega\beta}\right)^2}$$

$$-\dfrac{1}{Z}\dfrac{\partial Z}{\partial \beta} = \dfrac{\hbar\omega e^{-\hbar\omega\beta}}{1 - e^{-\hbar\omega\beta}} = \dfrac{\hbar\omega}{e^{\hbar\omega\beta} - 1}$$

$$U = \int\limits_0^\infty \dfrac{\hbar\omega}{e^{\hbar\omega\beta} - 1}\left(\dfrac{V\omega^2}{\pi^2 c^3}\right) d\omega$$

$$x = \hbar\omega\beta \qquad dx = \hbar\beta d\omega$$

$$U = \int\limits_0^\infty \dfrac{\hbar}{e^x - 1}\left(\dfrac{V}{\pi^2 c^3}\right)\left(\dfrac{x}{\hbar\beta}\right)^3 \dfrac{1}{\hbar\beta} dx = \dfrac{V}{\pi^2 c^3 \hbar^3 \beta^4}\int\limits_0^\infty \dfrac{x^3}{e^x - 1} dx = \left[\dfrac{k_B^4}{\pi^2 c^3 \hbar^3}\int\limits_0^\infty \dfrac{x^3}{e^x - 1} dx\right] V T^4$$

e. With $U = TS - PV$, $dU = TdS$ at constant V, let

$$a = \dfrac{k_B^4}{\pi^2 c^3 \hbar^3}\int\limits_0^\infty \dfrac{x^3}{e^x - 1} dx$$

$$\dfrac{dU}{T} = dS$$

$$\dfrac{4aVT^3 dT}{T} = dS$$

$$\dfrac{4}{3}aVT^3 = S$$

With $U = TS - PV$, $P = -\left(\dfrac{\partial U}{\partial V}\right)_S$,

$$U = aVT^4 = a\left(\dfrac{3}{4}\dfrac{S}{a}\right)^{\frac{4}{3}} V^{-\frac{1}{3}}$$

$$P = -\left(\frac{\partial U}{\partial V}\right)_S = \frac{1}{3}a\left(\frac{3}{4}\frac{S}{V}\right)^{\frac{4}{3}} = \frac{1}{3}aT^4$$

And from these results, clearly $U = TS - PV$ holds.

Solution 5.44

Problem 5.44

Throughout these solutions, let $\beta = \frac{1}{k_B T}$.

a. Using the canonical ensemble, we have (for one particle):

$$Z(\beta) = e^0 + e^{-\beta\varepsilon} + e^{-3\beta\varepsilon} + e^{-4\beta\varepsilon} + e^{-7\beta\varepsilon}$$

$$Z\left(\frac{2}{\varepsilon}\right) = 1 + e^{-2} + e^{-6} + e^{-8} + e^{-14}.$$

The proportion in state ε_4 is

$$P(\varepsilon_4, \beta) = \frac{e^{-8}}{Z\left(\frac{2}{\varepsilon}\right)},$$

The proportion in state ε_2 is

$$P(\varepsilon_2, \beta) = \frac{e^{-2}}{Z\left(\frac{2}{\varepsilon}\right)},$$

so that

$$\frac{P(\varepsilon_4, \beta)}{P(\varepsilon_2, \beta)} = \frac{e^{-8}}{e^{-2}} = e^{-6}.$$

b. Notice that the lowest ratio will be: $\dfrac{P(\varepsilon_5, \beta)}{P(\varepsilon_1, \beta)} = e^{-7\beta}$. Then want the temperature to be such

that $0.99 = -e^{-7\beta\varepsilon}$, or:

$$\ln 0.99 = -7\frac{1}{k_B T}\varepsilon$$

$$\ln 0.99 = \frac{1}{T}\frac{\varepsilon}{k_B}$$

$$T = -\frac{7}{\ln 0.99}\frac{\varepsilon}{k_B}.$$

c. The expected energy of a single particle is given by

$$E = \frac{\sum_i \varepsilon_i e^{-\beta\varepsilon_i}}{\sum_i e^{-\beta\varepsilon_i}} = -\frac{1}{Z}\frac{\partial Z}{\partial\beta} = \frac{\varepsilon e^{-\beta\varepsilon} + 3\varepsilon e^{-3\beta\varepsilon} + 4\varepsilon e^{-4\beta\varepsilon} + 7\varepsilon e^{-7\beta\varepsilon}}{e^0 + e^{-\beta\varepsilon} + e^{-3\beta\varepsilon} + e^{-4\beta\varepsilon} + e^{-7\beta\varepsilon}}$$

$$U = NE = N\frac{\varepsilon e^{-\beta\varepsilon} + 3\varepsilon e^{-3\beta\varepsilon} + 4\varepsilon e^{-4\beta\varepsilon} + 7\varepsilon e^{-7\beta\varepsilon}}{e^0 + e^{-\beta\varepsilon} + e^{-3\beta\varepsilon} + e^{-4\beta\varepsilon} + e^{-7\beta\varepsilon}}$$

At a very high temperature, the states are all equally populated for maximum entropy. This being the case,

$$U = NE = N\varepsilon\left(\frac{1}{5} + \frac{3}{5} + \frac{4}{5} + \frac{7}{5}\right) = 3N\varepsilon.$$

d. Consider momentarily the microcanonical ensemble, where $S = k_B\ln\Omega$, where omega counts the number of states. At this temperature, all of the particles are in the lowest quantum state and therefore the counting function gives one. $S = k_B\ln 1 = 0$.

e. Note that using the Helmholtz free energy, $A \equiv -\frac{1}{\beta}\ln Z$,

$$S \equiv \left(\frac{\partial A}{\partial T}\right)_{N,V} = -Nk_B\langle\ln P_i\rangle = -Nk_B\sum_i P_i\ln P_i.$$ At the highest temperature possible, each

state has equal occupation. In that case: $S = -Nk_B 5\left[\frac{1}{5}\ln\frac{1}{5}\right] = Nk_B\ln 5$.

f. $C_V = \left(\frac{\partial U}{\partial T}\right)_{N,V} = \left(\frac{\partial U}{\partial\beta}\right)_{N,V}\left(\frac{\partial\beta}{\partial\beta}\right) = -\frac{1}{k_B T^2}\left(\frac{\partial U}{\partial\beta}\right)_{N,V} = -k_B\beta^2\left(\frac{\partial U}{\partial\beta}\right)_{N,V}$

At very large temperatures, notice from part (c) that the internal energy is bounded above, and so at large temperatures

$$C_V = \left(\frac{\partial U}{\partial T}\right)_{N,V} \to 0.$$

Solution 5.45

a. The classical partition function is given for one atom by:

$$Z = \int_{-\infty}^{\infty} \int_{-\infty}^{\infty} e^{-\beta H(\vec{p},\vec{q})} d^n p\, d^n q$$

In this case:

$$Z = A \int_{-\infty}^{\infty} e^{-\beta\left(\frac{p^2}{2m} - \varepsilon_0\right)} d^n p = A \int_{0}^{\infty} e^{-\beta\left(\frac{r^2}{2m} - \varepsilon_0\right)} 2\pi r\, dr = 2\pi A e^{\beta\varepsilon_0} \int_{0}^{\infty} e^{-\beta\frac{r^2}{2m}} r\, dr$$

$$v = \beta\frac{r^2}{2m} \quad dv = \frac{\beta r}{m} dr \quad \int_{0}^{\infty} e^{-v} dv = 1$$

$$Z = 2\pi A \frac{m}{\beta} e^{\beta\varepsilon_0}$$

And so for N' particles:

$$Z_{N'} = \frac{1}{h^{2N'} N'!} \left(2\pi A \frac{m}{\beta} e^{\beta\varepsilon_0}\right)^{N'} = \frac{A^{N'} e^{N'\beta\varepsilon_0}}{N'!} \left(\frac{2\pi m}{h^2 \beta}\right)^{N'} = \frac{A^{N'} e^{N'\beta\varepsilon_0}}{N'!} \left(\frac{mkT}{2\pi\hbar^2}\right)^{N'}.$$

b. From the Helmholtz free energy $dF = -SdT - pdV + \mu dN$:

$$\mu = \left(\frac{\partial F}{\partial N}\right)_{T,V}. \text{ In this case,}$$

$$F = -\frac{1}{\beta}\ln Z = -\frac{1}{\beta}\left[-\ln N'! + N'\ln A + N'\ln\left(\frac{m}{2\pi\hbar^2\beta}\right) + N'\beta\varepsilon_0\right]$$

Use $\ln N! \approx N\ln N - N$:

$$\mu_S = \frac{\partial}{\partial N'}\left(-\frac{1}{\beta}\left[-\ln N'! + N'\ln A + N'\ln\left(\frac{m}{2\pi\hbar^2\beta}\right) + N'\beta\varepsilon_0\right]\right)$$

$$= \frac{\partial}{\partial N'} \left(-\frac{1}{\beta} \left[-N'\ln N' + N' + N'\ln A + N'\ln\left(\frac{m}{2\pi\hbar^2\beta}\right) + N'\beta\varepsilon_0 \right] \right)$$

$$= -\frac{1}{\beta}\ln\left(\frac{mA}{N'2\pi\hbar^2\beta}\right) - \varepsilon_0.$$

c. Again, $Z = \int\limits_{-\infty}^{\infty} \int\limits_{-\infty}^{\infty} e^{-\beta H\left(\vec{p},\vec{q}\right)}d^n q$. In 3 dimensions:

$$Z = V\int\limits_{-\infty}^{\infty} e^{-\beta\left(\frac{\vec{p}^2}{2m}\right)}d^n p = V\int\limits_{0}^{\infty} e^{-\beta\frac{r^2}{2m}}4\pi r^2 dr = 4\pi V\int\limits_{0}^{\infty} e^{-\beta\frac{r^2}{2m}}r^2 dr$$

$$u = \sqrt{\frac{\beta}{2m}}r \quad du = \sqrt{\frac{\beta}{2m}}dr$$

$$Z = 4\pi V\left(\frac{2m}{\beta}\right)^{\frac{3}{2}}\int\limits_{0}^{\infty} e^{-u^2}u^2 du = 4\pi V\left(\frac{2m}{\beta}\right)^{\frac{3}{2}}\left[\frac{\sqrt{\pi}}{4}\right] = V\left(\frac{2m\pi}{\beta}\right)^{\frac{3}{2}}$$

$$Z_{N-N'} = \frac{V^{N-N'}}{h^{3(N-N')}(N-N')!}\left(\frac{2m\pi}{\beta}\right)^{\frac{3}{2}(N-N')} = \frac{V^{N-N'}}{(N-N')!}\left(\frac{mkT}{2\pi\hbar^2}\right)^{\frac{3}{2}(N-N')}$$

d. Again, $\mu = \left(\frac{\partial F}{\partial N}\right)_{T,V}$.

$$F = -\frac{1}{\beta}\ln Z = -\frac{1}{\beta}\left[-\ln(N-N')! + (N-N')\ln V + \frac{3}{2}(N-N')\ln\left(\frac{m}{2\pi\hbar^2\beta}\right)\right]$$

$$\approx -\frac{1}{\beta}\left[-(N-N')\ln(N-N') + (N-N') + (N-N')\ln V + \frac{3}{2}(N-N')\ln\left(\frac{m}{2\pi\hbar^2\beta}\right)\right]$$

$$\mu_V = -\frac{1}{\beta}\left[-\ln(N-N') + \ln V + \frac{3}{2}\ln\left(\frac{m}{2\pi\hbar^2\beta}\right)\right] = -\frac{1}{\beta}\frac{3}{2}\ln\left[\left(\frac{V}{(N-N')}\right)^{\frac{2}{3}}\frac{m}{2\pi\hbar^2\beta}\right].$$

e. At equilibrium, $\mu_S = \mu_{S'}$.

$$-\frac{1}{\beta}\ln\left(\frac{mA}{N'2\pi\hbar^2\beta}\right) - \varepsilon_0 = -\frac{1}{\beta}\frac{3}{2}\ln\left[\left(\frac{V}{(N-N')}\right)^{\frac{2}{3}}\frac{m}{2\pi\hbar^2\beta}\right]$$

$$\frac{mA}{N'2\pi\hbar^2\beta}e^{\beta\varepsilon_0} = \frac{V}{(N-N')}\left(\frac{m}{2\pi\hbar^2\beta}\right)^{\frac{3}{2}}$$

$$\frac{N-N'}{N'} = \frac{V}{A}\left(\frac{m}{2\pi\hbar^2\beta}\right)^{\frac{1}{2}}e^{-\beta\varepsilon_0}.$$ This gives the ratio of atoms in the volume to atoms stuck to the surface.

f. In the limit of a very low temperature, β becomes infinite and the right hand side of the solution in part (e) becomes zero, indicating that the particles are all sticking to the walls. In the limit of a very high temperature, β goes to zero and the right hand side becomes infinite, indicating that all the particles are floating about the volume.

Solution 5.46

Problem 5.46

a. $N_{bulk} = (n-2)^3$, $N_{faces} = 6(n-2)^2$, $N_{edges} = 12(n-2)$, $N_{corners} = 8$

b. $E_{bulk} = \dfrac{-6b}{2}$, $E_{faces} = \dfrac{-5b}{2}$, $E_{edges} = \dfrac{-4b}{2}$, $E_{corners} = \dfrac{-3b}{2}$

c. Over a small volume range, the energy looks like this:

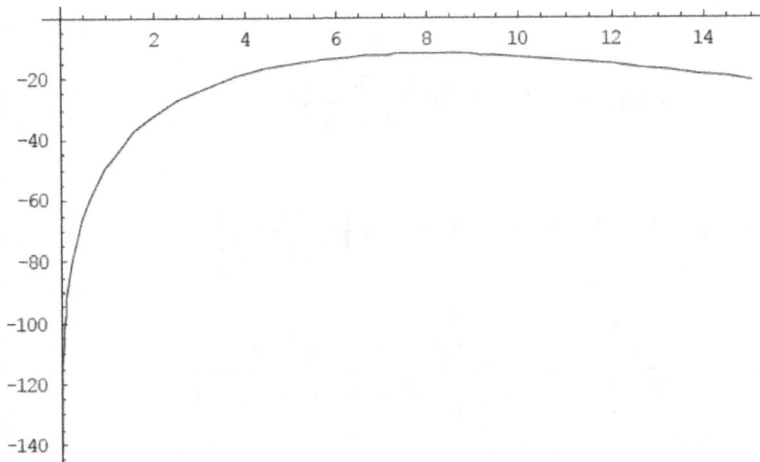

Over a larger volume range, it is almost linear:

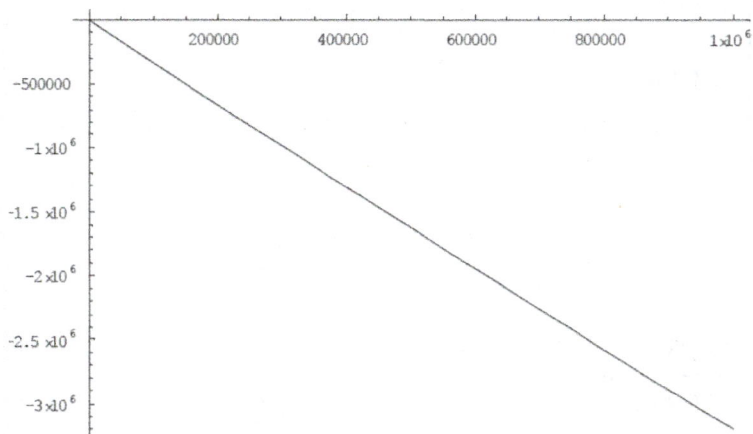

The proportion of the energy in the bulk actually grows relatively slowly. The volume would have to be on the order of 10^8 to have 99.9% of the energy in the bulk, as seen in the figure:

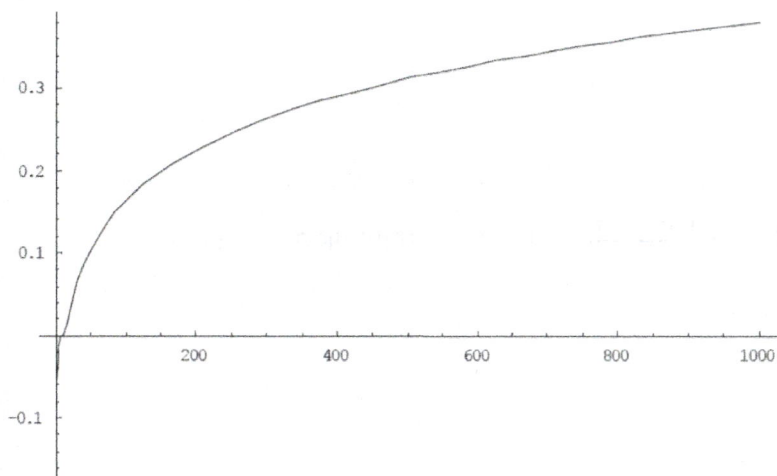

Solution 5.47

Problem 5.47

$$dG = -SdT + VdP + \mu dN$$

$$-\left(\frac{\partial S}{\partial P}\right)_{T,N} = \left(\frac{\partial V}{\partial T}\right)_{p,N} = \frac{\partial(T,S,N)}{\partial(p,T,N)}$$

$$\left(\frac{\partial T}{\partial p}\right)_{S,N} = \frac{\partial F(T,S,N)}{\partial F(p,T,N)}\frac{\partial F(p,T,N)}{\partial F(p,S,N)} = \frac{\partial F(V,p,N)}{\partial F(p,T,N)}\frac{\partial F(p,T,N)}{\partial F(p,S,N)}$$

$$= V \left[\frac{1}{V} \frac{\partial F(V,p,N)}{\partial F(p,T,N)} \right] \left[\frac{T}{T \frac{\partial F(p,S,N)}{\partial F(p,T,N)}} \right] = VT \frac{\alpha}{C_P}$$

For an ideal gas:

$$C_P = \frac{5}{2} nR \quad \alpha = \frac{1}{V} \left(\frac{\partial V}{\partial T} \right)_{p,N} = \frac{nR}{VP}$$

$$dT = \left(\frac{\partial T}{\partial p} \right)_{S,n} dP + \left(\frac{\partial T}{\partial S} \right)_{n,P} dS + \left(\frac{\partial T}{\partial n} \right)_{S,P} dn$$

$$dT = VT \frac{\alpha}{C_P} dP = \frac{2}{5} \frac{T}{P} dP$$

$$\frac{T}{P^{2/5}} = C = \frac{PV}{nRP^{2/5}} = \frac{P^{3/5}V}{nR}.$$

Solution 5.48

Problem 5.48

$$dG = -SdT + VdP + \mu dN$$

Using $-\left(\frac{\partial S}{\partial P} \right)_{T,N} = \left(\frac{\partial V}{\partial T} \right)_{p,N}$, get $\alpha_S = \frac{1}{V} \frac{\partial F(V,S,N)}{\partial F(V,p,N)}$ using the permutation rule. Now:

$$\frac{1}{C_V} \alpha_S = \frac{1}{TV} \frac{\partial F(T,V,N)}{\partial F(V,p,N)}$$

Giving

$$\frac{1}{C_V} \alpha_S = -\frac{1}{TV} \frac{\kappa_T}{\alpha} = -\frac{\alpha}{C_P - C_V} \Rightarrow \frac{\alpha_S}{\alpha} = \frac{1}{1-\gamma}.$$

For an ideal gas,

$$C_P = \frac{5}{2} nR \quad C_V = \frac{3}{2} nR \Rightarrow \frac{\alpha_S}{\alpha} = -\frac{3}{2}.$$

Solution 5.49

Problem 5.49

a. Let $P = \frac{p}{kT}$, $M = \frac{\mu}{kT}$; then

$$M^\alpha = \frac{P - A}{B} \quad (M^\gamma)^2 = \frac{P - C}{D}$$

Now:

$$d\mu = -sdT + vdP \quad v = \left(\frac{\partial \mu}{\partial p}\right)_T = \left(\frac{\partial M}{\partial p}\right)_T \Rightarrow v^\alpha = \frac{1}{B}, v^\gamma = \frac{1}{2DM^\gamma}.$$

b. In equilibrium, $p_I = p_{II}, T_I = T_{II}, \mu_I = \mu_{II}$ to force chemical potentials to be equal. Then we must solve:

$$\frac{P-C}{D} = \left(\frac{P-A}{B}\right)^2$$

Use the quadratic formula:

$$P = A + \frac{B^2}{2D} \pm \sqrt{\frac{B^2(A-C)}{D} + \left(\frac{B^2}{2D}\right)^2}, \text{ and so}$$

$$M^\alpha = M^\gamma = \frac{B}{2D} + \sqrt{\frac{(A-C)}{D} + \frac{B^2}{4D^2}}, \text{ where the positive sign is chosen because these values}$$

must be positive. Now it is straightforward to determine the pressure by substituting:

$$P = \frac{p}{kT} \quad M^\alpha = \frac{P-A}{B}.$$

Solution 5.50

Problem 5.50

There are only two energy sets that achieve this: One has energy zero, two have energy one, and one has energy two, and three with energy one and one with energy two. In the first, choose one particle to have energy zero and one to have energy two, for $3 \times 4 = 12$ ways. In the second, choose one to have energy two for 4 ways. Summing these, have $\Omega(5\varepsilon, 4) = 16$. If the particles are indistinguishable, there are still only two energy sets, which themselves are now indistinguishable so that $\Omega(5\varepsilon, 4) = 2$.

Solution 5.51

Problem 5.51

The fraction that lies within ΔR of the surface is:

$$\frac{\frac{\pi^{n/2}}{\left(\frac{n}{2}\right)!}(R+\Delta R)^n - \frac{\pi^{n/2}}{\left(\frac{n}{2}\right)!}(R)^n}{\frac{\pi^{n/2}}{\left(\frac{n}{2}\right)!}(R+\Delta R)^n} = 1 - \left(\frac{1}{1+\frac{\Delta R}{R}}\right)^n$$

$$1 - \left(\frac{1}{1 + \frac{\Delta R}{R}}\right)^n = 1 - \varepsilon \Rightarrow \frac{\Delta R}{R} = \frac{1}{\sqrt[n]{\varepsilon}} - 1$$

Solution 5.52

Problem 5.52

The number of distinct microstates are equal to the number of independent, positive-integer solutions to $n_x^2 + n_y^2 + n_z^2 = \frac{4V^{2/3}\varepsilon^2}{h^2 c^2} = \varepsilon^*$, and so $\sum_{r=1}^{3N} n_r^2 = \frac{4V^{2/3}E^2}{h^2 c^2} = E^*$. Now as usual, let the solutions to $\Sigma(E, V, N)$ count states of E less than or equal to the given energy, which correspond to the all-positive indexed lattice points inside a sphere. This gives:

$$\Sigma(N,V,E) \approx \left(\frac{1}{2}\right)^{3N} \left[\frac{\pi^{3N/2}}{\left(\frac{3N}{2}\right)!} E^{*3N/2}\right] = \left(\frac{\sqrt{4\pi}V^{1/3}E}{2hc}\right)^{3N} \left[\frac{1}{\left(\frac{3N}{2}\right)!}\right]$$

Using Stirling's approximation:

$$\ln\Sigma(N,V,E) \approx 3N\ln\left(\frac{\sqrt{4\pi}V^{1/3}E}{2hc}\right) - \frac{3N}{2}\ln\frac{3N}{2} + \frac{3N}{2} = 3N\ln\left(\frac{\sqrt{4\pi}V^{1/3}E}{2hc}\sqrt{\frac{2}{3N}}\right) + \frac{3N}{2}$$

Now $S(E, V, N) = k\ln\Sigma$, giving $E = V^{-1/3}\sqrt{\frac{3h^2 c^2 N}{2\pi}}\exp\left(-\frac{1}{2} + \frac{S}{3kN}\right)$. Now have:

$$T = \left(\frac{\partial E}{\partial S}\right)_{V,N} = \frac{hc\exp\left(-\frac{1}{2} + \frac{S}{3kN}\right)}{k\sqrt{6\pi N}V^{1/3}} = \frac{E}{3kN}\text{(substituting } S = k\ln\Sigma)$$

$$P = -\left(\frac{\partial E}{\partial V}\right)_{S,N} = \frac{E}{3V} = \frac{kNT}{V}, \text{ so the ideal gas law is satisfied. Now}$$

$$C_V = \left(\frac{\partial E}{\partial T}\right)_{V,N} = 3kN, \quad C_P = \left(\frac{\partial(E + PV)}{\partial T}\right)_{P,N} = 4kN \Rightarrow \gamma = \frac{4}{3}.$$

Solution 5.53

Problem 5.53

$$Z = \sum_{n=0}^{\infty} \exp\left[-\beta\hbar\omega\left(\left(n + \frac{1}{2}\right) - x\left(n + \frac{1}{2}\right)^2\right)\right] = \sum_{n=0}^{\infty}\left(\exp\left[-\beta\hbar\omega\left(n + \frac{1}{2}\right)\right]\right)^{\left(1 - x\left(n + \frac{1}{2}\right)\right)}$$

Defining $z_0 = \sum_{n=0}^{\infty} \exp\left[-\beta\hbar\omega\left(n+\frac{1}{2}\right)\right]$, expand in x:

$$Z \approx z_0 + ux\frac{\partial^2}{\partial u^2}z_0$$

$$A = -\frac{1}{\beta}\ln Z \approx \frac{-\hbar\omega}{u}\left[\ln z_0 + \frac{1}{z_0}ux\frac{\partial^2}{\partial u^2}z_0\right]$$

$$C = -T\frac{\partial^2 A}{\partial T^2} = \frac{\hbar\omega}{ku}\frac{-ku^2}{\hbar\omega}\frac{\partial}{\partial u}\frac{-ku^2}{\hbar\omega}\frac{\partial}{\partial u}\left[-\frac{\hbar\omega}{u}\left(\ln z_0 + \frac{1}{z_0}ux\frac{\partial^2}{\partial u^2}z_0\right)\right]$$

Evaluating the series gives the expected result.

References
Statistical and thermal physics

Original references to some of the problems in this chapter:

1. Ando, T., Matsumoto, Y. & Uemura, Y. Theory of Hall Effect in a Two-Dimensional Electron System. *J. Phys. Soc. Jpn.* **39** 279–288 (1975).

2. Boltzmann, L. Ableitung des Stefan'schen Gesetzes, betreffend die Abhängigkeit der Wärmestrahlung von der Temperatur aus der electromagnetischen Lichttheorie. *Annalen der Physik und Chemie* **22**, 291–294 (1884).

3. Boltzmann, L. *Lectures on Gas Theory*, (Dover Publications, New York, 1896).

4. Bose, S.N. Plancks Gesetz und Lichtquantenhypothese. *Zeitschrift für Physik* **26** (1924).

5. Chandrasekhar, S. The Maximum Mass of Ideal White Dwarfs. *Astrophysical Journal* **74**, 81-82 (1931).

6. Clapeyron, M.C. Mémoire sur la puissance motrice de la chaleur. *Journal de l'École polytechnique* **23**, 153–190 (1834).

7. Clausius, R. Ueber die bewegende Kraft der Wärme und die Gesetze, welche sich daraus für die Wärmelehre selbst ableiten lassen [On the motive power of heat and the laws which can be deduced therefrom regarding the theory of heat]... *Annalen der Physik* **155**, 500-524 (1850).

8. Einstein, A. Quantentheorie des einatomigen idealen Gases. *Sitzungsberichte der Preussischen Akademie der Wissenschaften* **1925**, 3-14 (1925).

9. Frenkel, J. On pre-breakdown phenomena in insulators and electronic semi-conductors. *Physical Review* **54**, 647-648 (1938).

10. Gibbs, J.W. *Elementary Principles in Statistical Mechanics, developed with especial reference to the rational foundation of thermodynamics*, (Charles Scribner's Sons, New York, 1902).

11. Kamerlingh Onnes, H. Further experiments with liquid helium. *Comm. Phys. Lab. Univ. Leiden* **120-122** (1911).

12. London, F. The λ-Phenomenon of Liquid Helium and the Bose–Einstein Degeneracy. *Nature* **141**, 643–644 (1938).

13. Stefan, J. Über die Beziehung zwischen der Wärmestrahlung und der Temperatur. *Sitzungsberichte der mathematisch-naturwissenschaftlichen Classe der kaiserlichen Akademie der Wissenschaften* **79**, 391-428 (1879).

14. van der Waals, J.D. Over de Continuiteit van den Gas- en Vloeistoftoestand (on the continuity of the gas and liquid state). PhD thesis, Leiden (1873).

15. van der Waals, J.D. The equation of state for gases and liquids. *Nobel Lectures in Physics*, 254–265 (1910).

Also see:

16. Chandler, D. *Introduction to Modern Statistical Mechanics*, (Oxford University Press, New York, 1987).

17. Wu, D., Chandler, D. & Chandler, D. *Solutions Manual for Introduction to Modern Statistical Mechanics*, (Oxford University Press, New York, 1988).

18. Huang, K. *Statistical Mechanics*, (Wiley, New York,, 1963).

19. McQuarrie, D.A. *Statistical Mechanics*, (University Science Books, Sausalito, Calif., 2000).

20. Kittel, C. & Kroemer, H. *Thermal Physics*, (W. H. Freeman, San Francisco, 1980).

21. Reif, F. *Fundamentals of Statistical and Thermal Physics,* (McGraw-Hill, New York, 1965).

22. Fowler, R.H. *Statistical Mechanics; the Theory of the Properties of Matter in Equilibrium*, (Cambridge U.P., Cambridge, 1966).

23. Irwin, P. *Giant Planets of Our Solar System: an Introduction*, (Springer; Praxis, Berlin; New York; Chichester, UK, 2006).

24. Hill, T.L. *An Introduction to Statistical Thermodynamics*, (Dover Publications, New York, 1986).

6

Electricity and magnetism

Problems

Electricity and magnetism

Electricity and magnetism is a huge field, and we don't pretend to be comprehensive here. We're trying to touch on the types of tricks you're likely to need on a final exam or qualifier: method of images, superpositions, etc. Our use of units isn't always consistent, so always beware of ε_0.

Problem 6.01

<u>Solution 6.01</u>

A conducting sphere (radius R_0) is placed into a uniform E field pointing in the z-direction. Find E everywhere.

Problem 6.02

<u>Solution 6.02</u>

A spherical shell of inner radius R_1 and outer radius R_2 is made of a material with polarization (dipole moment / volume) = P pointing in the z-direction. Find the electric field inside the hollow interior, within the shell, and outside the sphere.

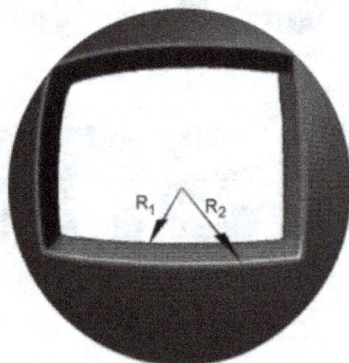

Problem 6.03

Solution 6.03

Now consider that the spherical shell is a charge distribution of charge density ρ. Calculate the electric field everywhere.

Problem 6.04

Solution 6.04

Consider a smaller sphere of radius R_1, charge density ρ_1, embedded within a larger sphere of radius R_2 and charge density ρ_2. Calculate:

a. The electric field

b. The potential inside the smaller sphere and outside both spheres

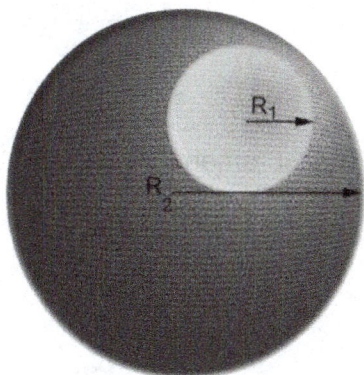

Problem 6.05

Solution 6.05

Now consider that the picture in Problem Problem 6.04 represents a conducting sphere with a hole in it, and that a point charge q is located at the center of the hole. Calculate the electric field inside the hole and outside the sphere.

Problem 6.06

Solution 6.06

A long iron rod with magnetic permeability μ is wound with wire (N turns/unit length). A current I flows through the wire. Calculate B and H within the rod.

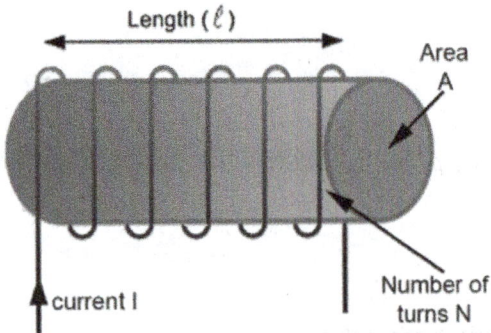

Problem 6.07

Solution 6.07

To determine the Stokes parameters of a beam of light, what measurements need to be made?

Problem 6.08

Solution 6.08

A plane wave of wavelength λ, moving in the z-direction, is incident at right angles to a dielectric film of thickness d and refractive index n . Calculate the fractions of incident energy that will be transmitted through/reflected by the film.

Problem 6.09

Solution 6.09

Negative hydrogen ions (H⁻), can be accelerated in a cyclotron, but an electric field of more than 1.5×10^4 gauss will pull off the extra electron. In order to accelerate H⁻ up to 1 GeV without losing the electron, what is the strongest B field that can be used? What is the radius of the resulting orbit?

Problem 6.10

Solution 6.10

Consider a parallel plate capacitor consisting of plates of radius R with separation d. If there is a uniform E field between the plates, perpendicular to them, calculate the energy contained in the electromagnetic field and the ratio of the energy in the B field to the energy in the E field.

Problem 6.11

Solution 6.11

Derive a formula for the differential scattering cross section of a circularly polarized plane wave by a free electron. If the incident plane wave is polarized with Stokes parameters P_x, P_y, P_z, what are the Stokes parameters of the scattered wave?

Problem 6.12

Solution 6.12

Derive the Larmor formula for radiation from an accelerated charge.

Problem 6.13

Solution 6.13

Calculate the radiated power per cycle for a charge q moving in an orbit $Z(t) = A\cos(\omega_0 t)$.

Problem 6.14

Solution 6.14

A dipole with charges $\pm q$ and diameter D rotates in a plane about the midpoint at constant speed. What is the angular distribution and polarization of the emitted radiation?

Problem 6.15

Solution 6.15

$2N$ point charges with a total charge q (each charge is $q/2N$) are arranged on a circle of radius R. If the charges move at a constant angular speed ω, what is the radiated power per unit solid angle (averaged over one cycle)? Discuss the special cases of $N = 2$ and $N \to \infty$.

Problem 6.16

Solution 6.16

Discuss the instability of the classical model of the hydrogen atom; that is, an electron of charge $-e$ moving in a stationary orbit around a proton of charge $+e$, with radius a_0. How long will it take for the electron to fall to the origin? Get a number in seconds and comment on its physical meaning.

Problem 6.17

Solution 6.17

A rotating conducting bar of length L is placed perpendicular to a magnetic field as shown, with one end fixed as indicated by the point. Calculate the induced EMF in the bar. See diagram on following page.

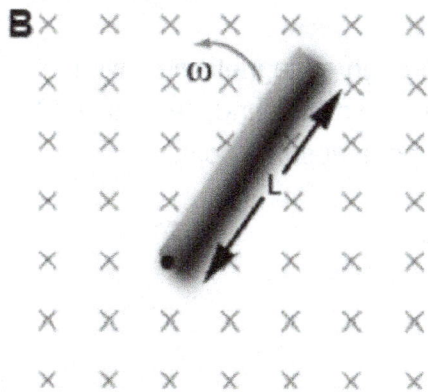

Problem 6.18

<u>Solution 6.18</u>

A circular loop of resistance r and radius a is placed in a time varying magnetic field, $B = B_0 + \omega t$. What is the induced EMF? The induced current?

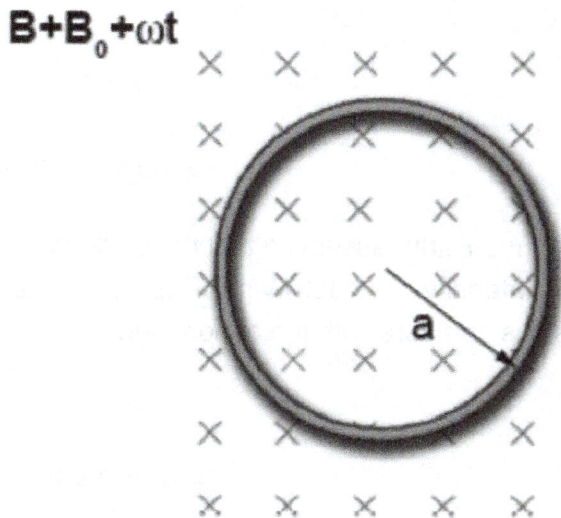

Problem 6.19

Solution 6.19

A square frame of resistivity r and side length a moves through a magnetic field with constant velocity v. What is the induced EMF as it exits the field? The induced current? If the frame is not accelerating as it moves, will it cease to move at some point?

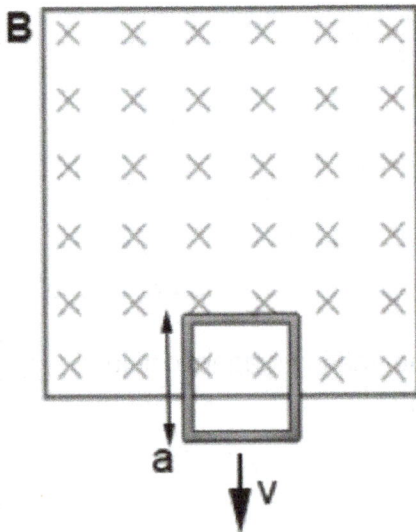

Problem 6.20

Solution 6.20

Calculate the radiation pressure on the surface of the Earth, assuming that the Earth is completely absorbing and that the solar radiation intensity is ~ 1300 W/m². Repeat for the more realistic case of 38% reflecting. How does this compare with the atmospheric pressure?

Problem 6.21

Solution 6.21

Poynting-Robertson effect. Due to radiation pressure, small particles above a critical size will eventually spiral into the Sun. Calculate the critical size at which this will happen for a spherical, completely absorbing particle orbiting the sun. What is an actual value, given a solar luminosity of $L = 4 \times 10^{26}$ watts? Do you imagine that this effect can be resolved for artificial satellites—and if so, which type would show the greatest effect?

Problem 6.22

Solution 6.22

For the pictured circuit, find V_{out} as a function of V_{in}.

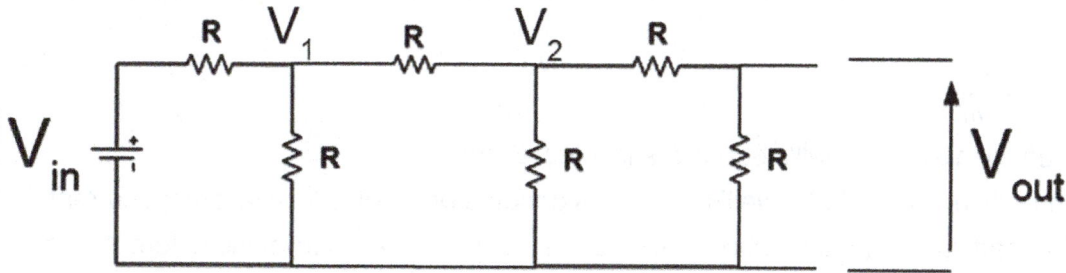

Problem 6.23

Solution 6.23

The Meissner effect says that for a superconductor, current flows frictionlessly, so the magnetic fields decay exponentially at the superconductor's surface, effectively "expelling" the field. Consider a superconducting sphere of radius a, placed in a magnetic field pointing uniformly in the z direction. Calculate the magnetic field at the surface of the sphere at $\theta = 0$.

Problem 6.24

Solution 6.24

A circle of conducting wire has a charge q and radius r. An electron is dropped from rest a distance x above the plane of the circle. With what speed does it pass through the circle? What is the condition for being able to treat this problem non-relativistically?

Problem 6.25

Solution 6.25

A charged particle of charge q is placed a distance d over a large, grounded conducting surface. Calculate the force on the particle and the distribution of charge on the surface.

Problem 6.26

Given the Maxwell's equations in differential form (SI units):

$$\vec{\nabla} \cdot \vec{D} = \rho$$

$$\vec{\nabla} \times \vec{E} + \frac{\partial \vec{B}}{\partial t} = 0$$

$$\vec{\nabla} \cdot \vec{B} = 0$$

$$\vec{\nabla} \times \vec{H} = \vec{J} + \frac{\partial \vec{D}}{\partial t},$$

a. Write down these Maxwell's equations in integral form.

b. Which term in which of Maxwell's equations denotes Maxwell's "displacement current?" For that particular equation, show how the integral form follows from the differential form.

c. Give an example (including a clearly labeled sketch) of a situation where omitting the displacement current term would lead to a glaring inconsistency. Explain clearly.

d.   i. Derive the wave equation for electromagnetic waves in a linear medium (characterized by ε and μ) in the absence of free charges and currents.

   ii. Write down, in complex notation, expressions describing the fields $\vec{E}$ and $\vec{B}$ of a plane electromagnetic wave propagating in a linear medium in the positive z direction.

   iii. How is the speed of propagation related to ε and μ?

e. Write down all the boundary conditions which electric and magnetic fields must satisfy when crossing the boundary from one linear medium (characterized by ε_1, μ_1) to another (ε_2, μ_2)? Pick one and show how it derives from the Maxwell's Equations.

f. When a plane electromagnetic wave, initially propagating in a vacuum, enters a linear medium of refraction index $n > 1$, which of the characteristic properties of the wave remain unchanged: wavelength, frequency, or speed?

g.   i. Derive expressions for the electric and magnetic field of a plane wave in an ohmic conductor of good conductivity in the absence of free charges.

   ii. Point out some characteristic differences in the behavior of these fields compared to the fields in the vacuum.

   iii. Derive an expression for the "skin depth" of a good ohmic conductor.

Problem 6.27

Solution 6.27

A solid glass sphere of mass M and radius R carries on its surface a uniformly distributed positive charge Q. It is spinning in outer space with angular speed ω.

a. Since the sphere is spinning, there are accelerated charges. Explain briefly why the spinning sphere does not emit electromagnetic radiation.

b. Next, an external magnetic field of magnetic field of magnitude B is established and the sphere moves in pure precession with the spin axis at an angle α to the magnetic field. Derive how much time it takes to precess through one complete turn.

Problem 6.28

Solution 6.28

An ideal circular parallel-plate capacitor of radius a and plate separation $d \ll a$ is connected to an alternating-current generator by axial leads, as shown in the figure. The current in the wire is $I(t) = I_0 \sin(\omega t)$, where $\omega \ll \frac{c}{a}$ and c is the speed of light. Assume that the axial leads lie along the z-axis with the origin at the center of the capacitor.

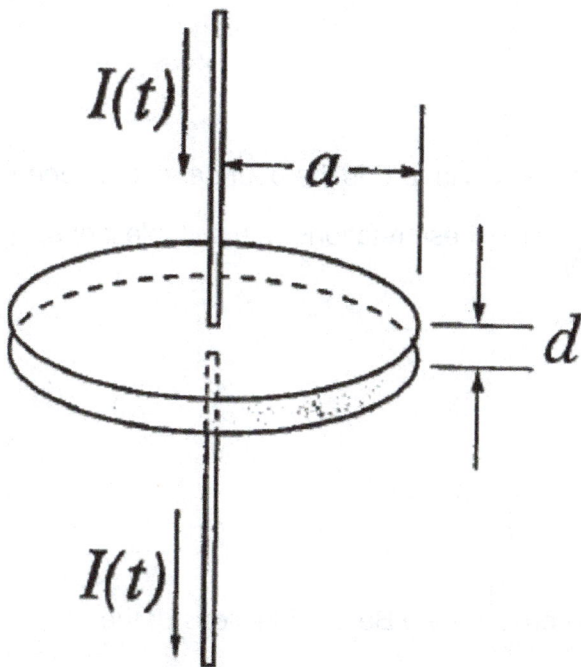

a. The electric and magnetic fields inside the capacitor between the plates have the form, neglecting fringe effects:

$$E(\rho,\phi,z) = E_0 f_E\left(\frac{\omega\rho}{c}\right)\cos(\omega t)\hat{k}$$

$$B(\rho,\phi,z) = -\frac{E_0}{c} f_B\left(\frac{\omega\rho}{c}\right)\sin(\omega t)\hat{e}_\phi,$$

where ρ,ϕ,z are standard cylindrical coordinates and E_0 is defined such that $f_E(0) = 1$. Note that the electric field at the center of the capacitor is given by $E_0\cos(\omega t)\hat{k}$. The functions $f_E(0), f_B(0)$, where $x = \frac{\omega\rho}{c}$, can be determined using the integral forms of Maxwell's equations:

$$\varepsilon_0 \int_{S_c} \vec{E} \cdot d\vec{a} = \int_V \rho dV$$

$$\oint_C \vec{E} \cdot d\vec{l} = -\frac{\partial}{\partial t}\int_{S_o} \vec{B} \cdot d\vec{a}$$

$$\varepsilon_0 \int_{S_c} \vec{B} \cdot d\vec{a} = 0$$

$$\frac{1}{\mu_0}\oint_C \vec{B} \cdot d\vec{l} = \int_{S_o} \vec{J} \cdot d\vec{a} + \varepsilon_0 \frac{\partial}{\partial t}\int_{S_o} \vec{E} \cdot d\vec{a}$$

where the closed surface S_c encloses volume V, closed loop C is the boundary of the open surface S_o, ρ is a charge density, and $c = \dfrac{1}{\sqrt{\mu_0\varepsilon_0}}$. Use these relations with suitable loops, surfaces, and volumes to show that:

$$f_E(x) = 1 - \int_0^x dx' f_B(x')$$

$$x f_B(x) = \int_0^x dx' x' f_E(x')$$

Next, show that $f_E(x) = J_0(x)$ and $f_B(x) = J_1(x)$, where $J_n(x)$ are Bessel functions of the first kind.

b. Show that neglecting fringe effects, the coefficient E_0 is given by

$$E_0 = \frac{I_0}{2\pi\varepsilon_0 acJ_1\left(\frac{\omega a}{c}\right)}.$$ You will need the following property of Bessel functions:

$$\frac{d}{dx}\left(x^n J_n(x)\right) = x^n J_{n-1}(x).$$

c. Use the small-x approximations $J_0(x) \approx 1 - \frac{1}{4}x^2$, $J_1(x) \approx \frac{1}{2}x - \frac{1}{16}x^3$ to calculate the capacitance C and the inductance L of the equivalent series LC circuit to leading order in $\frac{\omega a}{c}$. Note that with the above conventions for Maxwell's equations, the energy density of $\vec{E}$ is $\frac{1}{2}\varepsilon_0 \vec{E}^2$ and the energy density of $\vec{B}$ is $\frac{1}{2\mu_0}\vec{B}^2$. Given L and C, estimate the resonant frequency of the system in terms of $\frac{c}{a}$.

d. From a large distance away, the capacitor can be viewed simply as an oscillating electric dipole with $p_E(t) = \vec{p}_0\cos(\omega t)$. Calculate $\vec{p}_0$.

e. For a localized system of charges and current varying sinusoidally in time,

$$\rho\left(\vec{r},t\right) = \rho\left(\vec{r}\right)e^{-i\omega t}, \quad \vec{J}\left(\vec{r},t\right) = \vec{J}\left(\vec{r}\right)e^{-i\omega t},$$ the vector potential far from the capacitor

is given in the long-wavelength approximation and the Lorentz gauge by:

$$\vec{A}\left(\vec{r},t\right) = \vec{A}\left(\vec{r}\right)e^{-i\omega t}$$

$$\vec{A}(\vec{r},t) = \frac{e^{\frac{i\omega r}{c}}}{4\pi\varepsilon_0 c^2 r}\int d^3r\,\vec{J}(\vec{r}), \text{ where } r = \left|\vec{r}\right|.$$ Show that for an oscillating electric dipole:

$$\vec{A}(\vec{r}) = -\frac{i\omega\vec{p}_0}{4\pi\varepsilon_0 c^2}\frac{e^{\frac{i\omega r}{c}}}{r}.$$

f. For the magnetic field far from the capacitor, let $\vec{B}\left(\vec{r},t\right) = \vec{B}\left(\vec{r}\right)e^{-i\omega t}$, then use

$\vec{B} = \vec{\nabla}\times\vec{A}$ to show that:

$$\vec{B}(\vec{r}) = \frac{i\omega}{4\pi\varepsilon_0 c^2}\frac{\partial}{\partial r}\left(\frac{e^{\frac{i\omega r}{c}}}{r}\right)\vec{p}_0\times\hat{e}_r, \text{ where } \hat{e}_r = \frac{\vec{r}}{r}$$ is the unit vector directed radially away from

the origin. Determine the radiation field $\vec{B}_{\text{rad}}$ far from the capacitor and the time-averaged power $\langle P \rangle$ radiated, both in the long wavelength approximation. Note that the Poynting vector is $\vec{S} = \frac{1}{\mu_0}\vec{E}\times\vec{B}$.

Problem 6.29

Solution 6.29

a. A spherical shell of radius a has a surface charge density given in spherical coordinates by $\sigma = \sigma_0 \sin\theta\cos\phi$, where where σ_0 is a constant and the origin is at the center of the spherical shell. Find

$$Q = \int \rho d^3x$$

$$\vec{P} = \int \rho \vec{x} d^3x$$

$$Q_{i,j} = \int \rho\left(3x_i x_j - \delta_{i,j} x^2\right) d^3x$$

b. Assume some steady current density produces a magnetic field:

$$\vec{B} = \vec{\nabla} \times \vec{A}$$

$$\vec{A}(\vec{x},t) = \frac{\mu_0}{4\pi} \int \frac{\vec{J}(\vec{x}',t)}{|\vec{x} - \vec{x}'|} d^3x'$$

If we average the field over a sphere of radius a containing all of the current density:

$$\vec{B}_{avg} = \frac{3}{4\pi a^3} \int \vec{B}\, d^3x$$

Show that the average field is given as a function of the magnetic moment m as:

$$\vec{B}_{avg} = \frac{\mu_0}{4\pi} \frac{2\vec{m}}{a^3}.$$

Problem 6.30

Solution 6.30

Some dielectric material under some condition will have the property that the displacement D and the electric field E are related by $\vec{D} = \varepsilon_0 \vec{E} + i\vec{E} \times \vec{g}$, where g is a constant real vector. Consider plane waves propagating in the direction of g (considered to be the z direction), and starting from Maxwell's equations, find the possible values of the index of refraction n in terms of ε_0 and g.

Problem 6.31

Solution 6.31

Let x^μ and u^μ be the coordinates and the 4-velocity of a charged particle. Define the vector r^μ by $r^\mu = x^\mu - \frac{1}{c^2}(u \cdot x)r = u^\mu$.

a. Show that r^μ is space-like, i.e. $r^\mu r_\mu < 0$.

b. Let $r = \sqrt{r^\mu r_\mu}$ and consider the 4-vector potential A^μ of the form $A^\mu = \frac{qu^\mu}{4\pi\varepsilon_0 r}$. Compute the electric and magnetic fields from the 4-vector potential.

c. For the special case where the particle is at rest, what are the electric and magnetic fields from this formula?

Problem 6.32

Solution 6.32

For a point particle, the charge density and current density are given by:

$$\rho\left(\vec{x},t\right) = q\delta^3\left(\vec{x} - \vec{s}\,(t)\right)$$

$$\vec{J}\left(\vec{x},t\right) = q\vec{v}\,\delta^3\left(\vec{x} - \vec{s}\,(t)\right)'$$

where s is the position vector and v is the particle velocity. Show that the equation of continuity $\frac{\partial\rho}{\partial t} + \vec{\nabla}\cdot\vec{J} = 0$ is satisfied.

Problem 6.33

Solution 6.33

Prove the mean value theorem: for charge-free space, the value of the electric potential at any point is equal to the average of the potential over the surface of any sphere centered at that point.

Problem 6.34

Solution 6.34

An infinite straight wire carries the current $I(t) = \begin{cases} 0 & t \leq 0 \\ kt & t > 0 \end{cases}$

a. Find the retarded vector potential.

b. Compute the electric and magnetic fields.

Problem 6.35

Solution 6.35

Consider a wire loop of radius a with a current flowing through it of the form $I(t) = I_0\cos(\omega t)$. Suppose that $a \ll \dfrac{c}{\omega}$. Compute the electric and magnetic field in the zone $kr \gg 1$.

Problem 6.36

Solution 6.36

Consider two oppositely charged oscillating electric dipoles, separated by a distance d and oriented parallel to the direction connecting them.
a. Find the vector potential in the far zone to first order in d.
b. Compute the electric and magnetic fields in the same approximation.

Problem 6.37

Solution 6.37

A line charge of constant density λ and of length L lies in the first quadrant of the xy plane with one end at the origin. It makes an angle α with respect to the positive x-axis. Compute the dipole moment and all components of the quadrupole moments.

Problem 6.38

Solution 6.38

Suppose the vector potential is of the form $\vec{A} = \dfrac{1}{r^2}\left(\vec{a} \times \vec{r}\right) \times \vec{r}$. Compute the magnetic field.

Problem 6.39

Solution 6.39

A dipole p_1 is located at r_1 and another dipole p_2 is located at r_2.
a. Find the energy of this system.
b. Find the force on p_2 due to the electric field produced by p_1.
c. Suppose p_1 and p_2 are pointing in the same direction. Find the orientation of p_1 relative to $r_2 - r_1$ that minimizes the energy.

Problem 6.40

Solution 6.40

A uniformly charged solid sphere of radius R carries total charge Q and is set to spinning with angular velocity ω around the z-axis. It is given that for a spherical shell:

$$\vec{A}_{in} = \frac{\mu_0 a \omega \sigma}{3}(r\sin\theta)\,\widehat{\phi} \qquad \vec{A}_{out} = \frac{\mu_0 a^4 \omega \sigma}{3}\left(\frac{\sin\theta}{r^2}\right)\widehat{\phi}.$$

a. Find the magnetic field inside the sphere.

b. Find the magnetic moment of the sphere.

c. Find the average magnetic field over the sphere and express the average in terms of the magnetic moments.

Problem 6.41

Solution 6.41

Consider the relativistic motion of a charged particle in static and homogeneous electric and magnetic fields which are perpendicular to each other and where $\left|\vec{E}\right| = c\left|\vec{B}\right|$ in some inertial frame K.

a. Is it possible to find an inertial frame such that there is only magnetic field and no electric field? Explain.

b. If we go to a new inertial frame moving with velocity v in a direction perpendicular to E and B, find the fields in the new frame. Are they still perpendicular with $\left|\vec{E}\right| = c\left|\vec{B}\right|$?

c. Take B in the z-direction and E in the y-direction. Write down the equation of motion in the original frame K.

d. Show that p_z and $p_x - \dfrac{mc}{\sqrt{1 - \frac{v^2}{c^2}}}$ are conserved.

Problem 6.42

Consider the relativistic motion of a particle with charge *q* in a static Coulomb field described by the potential $\phi = \frac{e'}{r}$.

a. Show that the total energy $E = mc^2\gamma + q\phi$ is conserved.

b. Assuming that the particle motion is in a plane, use polar coordinates to find the motion.

Solutions
Electricity and magnetism

Solution 6.01

Problem 6.01

We need to solve the Laplace equation with the appropriate boundary conditions. Define the origin at the center of the sphere. Now we have:

$$\nabla^2 \Phi = 0$$

$$\Phi(\infty) = -Ez = -Er\cos\theta$$

$$\Phi(R_0) = \text{constant}$$

A general solution with the appropriate symmetry is:

$$\Phi(r,\theta,\phi) = \left[\sum_\ell a_\ell r^\ell + \frac{b_\ell}{r^{\ell+1}} \right] P_\ell(\cos\theta)$$

Applying the first boundary condition gives:

$$\Phi(r,\theta,\varphi) = -ErP_1(\cos\theta) + \sum_\ell \frac{b_\ell}{r^{\ell+1}}(\cos\theta)$$

Applying the second boundary condition:

$$-ER_0 + \frac{b_1}{R_0^2} = 0$$

$$\Phi(r,\theta,\varphi)EP_1(\cos\theta)\left[\frac{R_0^3}{r^2} - r \right] + \frac{b_0}{r}$$

$$\vec{E} = -\nabla\Phi = -E\left[\left(\frac{R_0^3}{r^3} \right)\hat{z} - 3\frac{R_0^3}{r^4}z\hat{r} \right] + \frac{b_0}{r^2}\hat{r}$$

Solution 6.02

We can think of the spherical shell as a superposition of a sphere of radius R_2 and polarization P, and a second sphere of radius R_1 and polarization $-P$. For a solid sphere of polarization P and radius R:

$$\vec{E} = \frac{4}{3}\pi P \hat{z} \text{ for } r < R$$

$$\vec{E} = \frac{4}{3}\pi P R^3 \left[\frac{\hat{z}}{r^3} - \frac{3z\hat{r}}{r^4} \right] \text{ for } r > R$$

Then the superposition gives:

$$\vec{E} = 0 \text{ for } r < R_1$$

$$\vec{E} = -\frac{4}{3}\pi P \left[\left(1 - \frac{R_1^3}{r^3} \right)\hat{z} - \frac{3zR_1^3\,\vec{r}}{r^4} \right] \text{ for } R_1 < r < R_2$$

$$\vec{E} = -\frac{4}{3}\pi P \left(R_2^3 - R_1^3 \right) \left[\frac{\hat{z}}{r^3} - \frac{3z\hat{r}}{r^4} \right] \text{ for } r > R_2$$

Solution 6.03

The field is radial and spherically symmetric.

$E = 0$ for $r < R_1$ since no charge is enclosed

$\vec{E} = \frac{Q}{r^2}\hat{r}$ for $r > R_2$ where Q is the total charge enclosed

For $R_1 < r < R_2$:

$$E(r) = \frac{4\pi}{r^2} \int_{R_1}^{r} \rho(r')r'^2 dr'$$

Solution 6.04

Consider this as a superposition of two uniformly charged spheres.

a. Recall that for the case of a uniformly charged sphere of radius R and total charge Q:

$$\vec{E} = \frac{Q}{r^2}\hat{r} \text{ for } r > R$$

$$\vec{E} = \frac{Q}{r^2}\left(\frac{r}{R}\right)^3\hat{r} = \frac{Q}{R^3}r\hat{r} \text{ for } r < R$$

where the latter case takes into account the charge contained within the radius r. To take the superposition, need to define a vector that connects the center of the big sphere with the center of the small one; call this r'. Then we have:

$$\vec{E} = \frac{4}{3}\pi(\rho_1 - \rho_2)\vec{r} + \frac{4}{3}\pi\rho_2\left(\vec{r} + \vec{r}'\right) \text{ for } r < R_1$$

$$\vec{E} = \frac{4}{3}\pi R_1^3(\rho_1 - \rho_2)\frac{\vec{r}}{r^3} + \frac{4}{3}\pi R_2^3\rho_2\frac{\vec{r} + \vec{r}'}{\left|\vec{r} + \vec{r}'\right|^3} \text{ for } r > R_2$$

b. Outside a uniformly charged sphere of radius R and total charge Q:

$$V = \frac{Q}{r}$$

Inside the sphere:

$$-\frac{\partial V}{\partial r} = E(r) = \frac{Q}{R^3}r$$

$$V(r) = -\frac{Q}{2R^3}r^2 + \text{const}$$

Choose the constant to make the potential continuous at $r = R$, to give:

$$V(r) = \frac{1}{2}\frac{Q}{R}\left[3 - \frac{r^2}{R^2}\right]$$

Applying this to our superposition gives:

$$V = \frac{1}{2}\frac{4}{3}\pi R_1^2 (\rho_1 - \rho_2) \left[3 - \frac{r^2}{R_1^2} \right] + \frac{1}{2}\frac{4}{3}\pi R_2^2 \left[3 - \frac{\left| \vec{r} + \vec{r}' \right|^2}{R_2^2} \right] \text{ for } r < R_1$$

$$V = \frac{4}{3}\pi R_1^3 (\rho_1 - \rho_2) \frac{1}{r} + \frac{4}{3}\pi R_2^3 \rho_2 \frac{1}{\left| \vec{r} + \vec{r}' \right|} \text{ for } r > R_2$$

Solution 6.05

Problem 6.05

The potential is given by:

$\nabla^2 V = -4\pi q$ inside the hole

$\nabla^2 V = -0$ outside the hole

We can solve this by keeping $V(r)$ constant at R_1 and R_2, to give:

$\vec{E} = -\nabla V = \frac{q}{r^2}\hat{r}$ inside the hole

$\vec{E} = \frac{q}{\left| \vec{r} + \vec{r}' \right|^3}\left(\vec{r} + \vec{r}' \right)$ outside the sphere

Solution 6.06

Problem 6.06

Use:

$$\text{curl}\vec{H} = \frac{4\pi}{c}\vec{J}_{\text{free}}$$

$\int \text{curl}\vec{H} \cdot d\vec{A} = \oint \vec{H} \cdot d\vec{l} = \frac{4\pi}{c}In$, where n = number of turns crossing the area. Integrating over the length allows us to cancel the length:

$$Hl = \frac{4\pi}{c} INl$$

$$H = \frac{4\pi}{c} IN$$

$$B = \frac{4\pi}{c} \mu IN$$

Solution 6.07

Problem 6.07

We want to define the matrix:

$$\rho = \frac{1}{2} \begin{pmatrix} 1 + P_z & P_x - iP_y \\ P_x + iP_y & 1 - P_z \end{pmatrix}$$

a. Measure intensity (using a light meter).

b. Get a polarized sheet marked with a line to determine the direction of the light coming out. Define x and y as perpendicular to each other and to the propagation. Align the sheet with x, measure the intensity of transmitted light, and divide by the intensity of incident light to obtain:

$$\sum_{\alpha,\beta} n_\alpha^* \rho_{\alpha\beta} n_\beta \text{ with } n = x, \text{ or } \rho_{xx} = \frac{1}{2}(1 + P_z)$$

c. Align the sheet midway between x and y. Repeat the experiment and set the value equal to:

$$\sum_{\alpha,\beta} n_\alpha^* \rho_{\alpha\beta} n_\beta \text{ with } \hat{n} = \frac{\hat{x} + \hat{y}}{\sqrt{2}}, \text{ or}$$

$$\frac{1}{2}\left(1 + \rho_{xy} + \rho_{yx}\right)$$

$$\rho_{xy} + \rho_{yx} = P_z$$

d. Now we need a device that passes only right-handed (left-handed) circularly polarized light. Use this to measure the fraction of transmitted light, and set that equal to:

$$\sum_{\alpha,\beta} n_\alpha^* \rho_{\alpha\beta} n_\beta \text{ with } \hat{n} = \frac{\hat{x} + i\hat{y}}{\sqrt{2}}, \text{ or}$$

$$\frac{1}{2}\left(1 + \left[\rho_{xy} - \rho_{yx}\right]\right)$$

$$i\left[\rho_{xy} - \rho_{yx}\right] = P_y$$

Solution 6.08

Problem 6.08

Assume the film is located between $z = 0$ and d. The parallel components E and B (assuming $\mu = 1$) will be continuous at $z = 0$ and $z = d$. Write the equations for E and B in each region and set them continuous at the boundaries.

For $z < 0$:

$$\vec{E} = E_i \hat{x} \exp\left(2\pi i \left[\frac{z}{\lambda} - vt\right]\right) + E_r \hat{x} \exp\left(2\pi i \left[-\frac{z}{\lambda} - vt\right]\right)$$

$$\vec{B} = E_i \hat{y} \exp\left(2\pi i \left[\frac{z}{\lambda} - vt\right]\right) - E_r \hat{y} \exp\left(2\pi i \left[-\frac{z}{\lambda} - vt\right]\right)$$

For $0 < z < d$:

$$\vec{E} = E_1 \hat{x} \exp\left(2\pi i \left[n\frac{z}{\lambda} - vt\right]\right) + E_2 \hat{x} \exp\left(2\pi i \left[-n\frac{z}{\lambda} - vt\right]\right)$$

$$\vec{B} = nE_1 \hat{y} \exp\left(2\pi i \left[\frac{z}{\lambda} - vt\right]\right) + E_2 \hat{y} \exp\left(2\pi i \left[-n\frac{z}{\lambda} - vt\right]\right)$$

For $z > d$:

$$\vec{E} = E_t \hat{x} \exp\left(2\pi i \left[\frac{z}{\lambda} - vt\right]\right)$$

$$\vec{B} = E_t \hat{y} \exp\left(2\pi i \left[\frac{z}{\lambda} - vt\right]\right)$$

Applying the boundary conditions gives four simultaneous equations for five unknowns; express everything as fraction of E_i:

$$E_1 + E_2 - E_r = E_i$$

$$nE_1 + nE_2 + E_r = E_i$$

$$x^n E_1 + x^{-n} E_2 - xE_t = 0$$

$$nx^n E_1 + nx^{-n} E_2 - xE_t = 0$$

Where:

$$x \equiv \exp\left(\frac{2\pi i d}{\lambda}\right)$$

Solve for E_r and E_t as functions of E_i:

$$E_r = \frac{\begin{vmatrix} 1 & 1 & 1 & 0 \\ n & -n & 1 & 0 \\ x^n & x^{-n} & 0 & -x \\ nx^n & -nx^{-n} & 0 & -x \end{vmatrix}}{\begin{vmatrix} 1 & 1 & -1 & 0 \\ n & -n & 1 & 0 \\ x^n & x^{-n} & 0 & -x \\ nx^n & -nx^{-n} & 0 & -x \end{vmatrix}} E_i = \frac{n^2 - 1}{n^2 + 1 + 2ni\cot\left(\frac{2\pi nd}{\lambda}\right)} E_i$$

The fraction of energy reflected is given by:

$$r = \left|\frac{E_r}{E_i}\right|^2 = \frac{(n^2 + 1)^2}{(n^2 - 1)^2 + 4n^2\cot^2\left(\frac{2\pi nd}{\lambda}\right)}$$

The fraction transmitted may be calculated similarly, or simply by using $t = 1 - r$.

Solution 6.09

For a charged particle moving in a magnetic field B, there will be an electric field:

$$\vec{E} = \gamma \frac{\vec{v} \times \vec{B}}{c} = \frac{\gamma vB}{c} \text{ for } \vec{v} \perp \vec{B}$$

So we require:

$$\frac{\gamma vB}{c} < 1.5 \times 10^4 G$$

$$B < \left(1.5 \times 10^4 G\right) \frac{c}{\gamma v}$$

If the kinetic energy of the ion is T, then use:

$$T + mc^2 = \gamma mc^2$$

$$\gamma = \frac{T}{mc^2} + 1 = \frac{1}{\sqrt{1 - \frac{v^2}{c^2}}}$$

$$\left(\frac{\gamma v}{c}\right)^2 = \gamma^2 - 1 = \left(\frac{T}{mc^2}\right)^2 + 2\frac{T}{mc^2}$$

which gives

$$B < \frac{\left(1.5 \times 10^4 G\right)}{\sqrt{\left(\frac{T}{mc^2}\right)^2 + 2\frac{T}{mc^2}}} = 0.866T$$

The radius of the orbit r is given by:

$$\omega_c = \frac{eB}{\gamma mc} = \frac{v}{r}$$

Solve for r, plug in values to get $r = 1000/1.5$ cm.

Solution 6.10

Use:

$$curl\vec{B} - \frac{1}{c}\frac{\partial E}{\partial t} = 0$$

$$\int curl\vec{B} \cdot d\vec{A} = \frac{1}{c}\int \frac{\partial E}{\partial t} \cdot d\vec{A} = \oint \vec{B} \cdot d\vec{l}$$

Take the area of integration to be a concentric circle between the plates of radius r; then:

$$d\vec{A} = \hat{z}dA$$

$$d\vec{l} = \hat{\theta}rd\theta$$

$$\vec{E} = E\hat{z}$$

and

$$B(2\pi r) = \frac{\pi r^2}{c}\frac{\partial E}{\partial t}$$

$$E^2 + B^2 = E^2 + \left(\frac{r}{2c}\frac{\partial E}{\partial t}\right)^2$$

Now plug into:

$$U = \int u dv = \frac{1}{8\pi}(E^2 + B^2)dv$$

$$= \frac{1}{8\pi}\left[\pi R^2 E^2 d + \frac{d}{4c^2}\left(\frac{\partial E}{\partial t}\right)^2 \int_0^R 2\pi r^3 dr\right]$$

$$= \frac{R^2 d}{8}\left[E^2 + \frac{R^2}{8c^2}\left(\frac{\partial E}{\partial t}\right)^2\right]$$

$$\frac{\text{energy in } B \text{ field}}{\text{energy in } E \text{ field}} = \frac{1}{E^2}\frac{R^2}{8c^2}\left(\frac{\partial E}{\partial t}\right)^2$$

Solution 6.11

$$\vec{E}_{\text{incident}} = E_0 \frac{\hat{x} \pm \hat{y}}{\sqrt{2}} \exp(ikz - i\omega t)$$

$$\vec{B}_{\text{incident}} = \hat{z} \times \vec{E}_{\text{incident}} = E_0 \frac{\hat{y} \mp i\hat{x}}{\sqrt{2}} \exp(ikz - i\omega t)$$

$$\vec{S}_{\text{incident}} = \frac{c}{8\pi} \text{Re}\left[\vec{E}_{\text{incident}} \times \vec{B}_{\text{incident}}^*\right] = \frac{c}{8\pi} E_0^2 \hat{z}$$

$$m\vec{a} = e\vec{E}_{\text{incident}} = eE_0 \frac{\hat{x} \pm \hat{y}}{\sqrt{2}} \exp(ikz - i\omega t)$$

$$E_{\text{rad}} = \frac{e}{e^2 R}\left(\hat{R} \times \left(\hat{R} \times \vec{a}\right)\right) = \frac{e^2 E_0}{\sqrt{2} mc^2 R}$$

$$S_{\text{rad}} = \frac{c}{8\pi} \vec{E}_{\text{rad}} \cdot \vec{E}_{\text{rad}}^* = \frac{c}{8\pi} \frac{e^4 E_0^2}{2m^2 c^4 R^2}\left(\hat{R} \times \left(\hat{R} \times \left[\hat{x} \pm \hat{y}\right]\right)\right) \cdot \left(\hat{R} \times \left(\hat{R} \times \left[\hat{x} \mp \hat{y}\right]\right)\right)$$

Remembering the formula:

$$\left[\vec{a} \times \vec{b}\right] \cdot \left[\vec{c} \times \vec{d}\right] = \left[\vec{a} \cdot \vec{c}\right]\left[\vec{b} \cdot \vec{d}\right] - \left[\vec{a} \cdot \vec{d}\right]\left[\vec{b} \cdot \vec{c}\right]$$

Gives:

$$\left(\hat{R} \times \left(\hat{R} \times \left[\hat{x} \pm \hat{y}\right]\right)\right) \cdot \left(\hat{R} \times \left(\hat{R} \times \left[\hat{x} \mp \hat{y}\right]\right)\right) =$$

$$\hat{R} \cdot \hat{R}\left[\hat{x} \pm \hat{y}\right]\left[\hat{x} \mp \hat{y}\right] - \left(\hat{R} \cdot \left[\hat{x} \mp \hat{y}\right]\right)\left(\left[\hat{x} \pm \hat{y}\right] \cdot \hat{R}\right) = 2 - \sin^2\theta$$

Giving:

$$S_{\text{rad}} = \frac{c}{8\pi} \frac{e^4 E_0^2}{2m^2 c^4 R^2} \left(2 - \sin^2\theta\right)$$

$$\frac{d\sigma}{d\Omega} = R^2 \frac{S_{\text{rad}}}{S_{\text{inc}}} = \frac{e^4}{m^2 c^4} \left(2 - \sin^2\theta\right)$$

Solution 6.12

Take a charge q with radial field lines and accelerate it to a velocity dv in time dt. The field lines around the charge in its new position will remain radial, causing a "kink" with an angular component of the E field appearing after time t: $\dfrac{E_0}{E_r} = \dfrac{t\sin\theta}{c}\dfrac{dv}{dt} = \dfrac{r\sin\theta}{c}\dfrac{dv}{dt}$.

From Coulomb's law:

$$E_r = \frac{q}{r^2} \Rightarrow E_\theta = \frac{qa\sin\theta}{rc^2}$$

The power radiated per unit area is given by the Poynting flux:

$$\vec{S} = \frac{c}{4\pi} E_0^2 \hat{k} = \frac{1}{4\pi} \frac{q^2 a^2 \sin^2\theta}{r^2 c^3}$$

Then the total power is the integral of the Poynting flux over a sphere:

$$P = \frac{1}{4\pi} \frac{q^2 a^2}{c^3} \int_0^\pi d\phi \int_0^\pi d\theta \frac{\sin^2\theta}{r^2} r^2 dr = \frac{2}{3} \frac{q^2 a^2}{c^3}$$

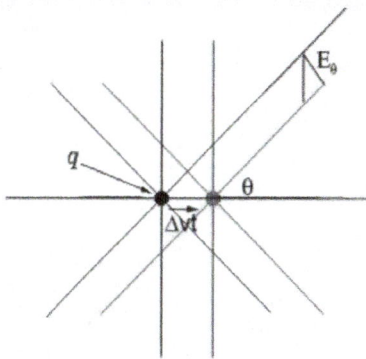

Solution 6.13

$$\frac{dP}{d\Omega} = \frac{e^2 \overline{\vec{a}^2}}{4\pi c^3} \sin^2\theta$$

where

a = acceleration of the particle, $\overline{\vec{a}}$ = time average, θ = angle between $\hat{z}$ and $\hat{R}$

$$\vec{a} = \frac{d^2 z}{dt^2} = -A\omega_0^2 \cos(\omega_0 t)$$

Average over a cycle:

$$\frac{dP}{d\Omega} = \frac{e^2 A^2 \omega_0^4}{8\pi c^3} \sin^2\theta.$$

Solution 6.14

$$\frac{dP}{d\Omega} = \frac{e^2 \overline{\vec{a}^2}}{4\pi c^3}$$

$$\vec{a} = \hat{R} \times \left(\hat{R} \times \left[\frac{d^2 \vec{x}_1}{dt^2} - \frac{d^2 \vec{x}_2}{dt^2} \right] \right)$$

$$x_{1,2}(t) = \pm \frac{D}{2}\left(\hat{x}\cos\omega t + \hat{y}\sin\omega t \right)$$

$$\frac{d^2 x_{1,2}}{dt^2} = \mp \frac{D}{2}\left(\hat{x}\cos\omega t + \hat{y}\sin\omega t \right)$$

Note that:

$$\left| \hat{R} \times \left(\hat{R} \times \hat{x} \right) \right|^2 = 1 - \left(\hat{R} \cdot \hat{x} \right)^2 = 1 - \sin^2\theta\cos^2\phi$$

$$\left| \hat{R} \times \left(\hat{R} \times \hat{y} \right) \right|^2 = 1 - \left(\hat{R} \cdot \hat{y} \right)^2 = 1 - \sin^2\theta\sin^2\phi$$

Plug this into the first equation to get:

$$\frac{dP}{d\Omega} = \frac{e^2 D^2 \omega^4}{8\pi c^3}(2 - \sin^2\theta) = \frac{e^2 D^2 \omega^4}{8\pi c^3}(1 + \cos^2\theta)$$

Solution 6.15

Problem 6.15

The charge density can be expressed as a function of the charge distribution and speed of rotation as:

$$\rho = \frac{q}{2N} \sum_{2N}^{i=1} \delta\left(\phi - \omega t - \frac{2\pi}{2N} i \right)$$

This is a periodic function, so it can be expanded in a Fourier series:

$$\rho(\phi,t) = \sum_{\infty}^{j=0} \rho_j(\phi) \exp\left(-ij[2N\omega]t \right)$$

where

$$\rho_j(\phi) = \frac{1}{2\pi} \int_0^{2\pi} d\left([2N\omega]t \right) \exp\left(ij[2N\omega]t \right) \rho(\phi,t) = \frac{1}{2\pi} \frac{q}{2N} \frac{2N\omega}{\omega} \sum_N^{i=1} \int_0^{\pi/N} \delta\left(\phi - \omega t - \frac{\pi}{N} i \right)$$

The only term that does not vanish is $i\frac{\pi}{N} \leq \phi \leq (i+1)\frac{\pi}{N}$,

$$\Rightarrow \rho_j(\phi) = \frac{q}{2\pi} \exp(2ijN\phi)$$

For two charges only ($N = 1$), the dipole radiation vanishes because:

$$\vec{P} = \frac{qR}{2\pi} \int_0^{2\pi} \exp(2ij\phi) \left[\hat{x}\cos\phi + \hat{y}\sin\phi \right] = 0;$$ thus, the lowest level will be electric quadrupole radiation with:

$$\vec{A}(\vec{r}) = \frac{1}{2c}\exp(-2i\omega t)\frac{k}{ir^2}\exp(ikr) \int J(\vec{r}')\vec{r} \cdot \vec{r}'d^3\vec{r}'$$

We can calculate the vector potential from the current density:

$$\vec{J}_j(\phi) = \frac{qv}{2\pi} \left[\hat{y}\cos\phi - \hat{x}\sin\phi \right]$$

Giving:

$$\int J_j(\phi')\vec{r} \cdot \vec{r}'d^3\vec{r}' = \frac{-iqva}{4}(x + iy)\left(\hat{x} + i\hat{y} \right)$$

We can get the magnetic field and Poynting vector:

$$\vec{B} = curl\vec{A}(\vec{r}) = \frac{1}{8c}\exp(-2i\omega t)\frac{kqva}{r^2}(x + iy)\vec{k} \times \left(\hat{x} + i\hat{y} \right)$$

$$\vec{S} = \frac{c}{8\pi}\frac{1}{r^2}\left(\frac{k^2qva}{8c} \right)^2 \left| (x + iy)\vec{k} \times \left(\hat{x} + i\hat{y} \right) \right|^2$$

$$\frac{dP}{d\Omega} = \frac{1}{512\pi c}(k^2qva)^2 \left| \vec{k} \times \left(\hat{x} + i\hat{y} \right) \right|^2 \frac{1}{512\pi c}(k^2qva)^2(2 - \sin^2\theta)$$

Substitute $v = a\omega = 2akc$ to get:

$$\frac{dP}{d\Omega} = \frac{c}{128}q^2a^4k^6(1 + \cos^2\theta)$$

For four charges, only the octupole term survives. When $N \to \infty$, there is no radiation.

Solution 6.16 Problem 6.16

$$\frac{dU}{dt} = \frac{-2e^2a^2}{3c^2}$$

where a is the acceleration of the electron. If the orbit is assumed to be circular, then:

$$\frac{dU}{dt} = \frac{-2e^2}{3r^4m^2c^3} = -\frac{2}{3}\frac{r_0^3c^3}{r^4}$$

where $r_0 \equiv \frac{e^2}{mc^2}$.

The total classical energy (kinetic plus potential) is given by:

$$U = \frac{1}{2}mv^2 - \frac{e^2}{r} = -\frac{r_0}{r}mc^2$$

$$\Rightarrow \frac{dU}{dt} = \frac{r_0}{2r^2}mc^2\frac{dr}{dt} = -\frac{2}{3}\frac{r_0^3}{r^4}mc^3$$

$$r^2\frac{dr}{dt} = \frac{1}{3}\frac{d}{dt}\left(r^3\right) = -\frac{4}{3}cr_0^2$$

which gives upon integration from 0 to a_0:

$$r^3 = a_0^3 - 4cta_0^3 \equiv 0 \text{ at } t = t_{critical}$$

$$t_{critical} = \frac{a_0^3}{4cr_0^2}.$$

This works out to about 10^{-11} seconds, which is actually representative of the lifetime of excited states of hydrogen.

Solution 6.17 Problem 6.17

Each small element of the bar, dr, a distance r from the pivot point will have an induced EMF:

$$d\varepsilon = Br\omega dr$$

Then integrate along the length:

$$\int d\varepsilon = B\int_0^L r\omega dr = \frac{1}{2}B\omega r^2$$

Solution 6.18

Problem 6.18

By Faraday's law, the induced EMF is given by the derivative of the flux:

$$\varepsilon = -\frac{d\Phi_B}{dt} = \frac{d}{dt}(BA) = -\pi a^2 \frac{dB}{dt} = -\pi a^2 \omega$$

Then the induced current is:

$$I = \frac{|\varepsilon|}{r} = \frac{\pi \omega a^2}{r}$$

Solution 6.19

Problem 6.19

The flux through the frame decreases at the rate Bva, so the induced EMF is $(1/c)Bva$. The total resistance of the frame is $4ar/A$, where A is the cross-sectional area of the sides. This gives an induced current of $I = $ EMF/total resistance $= BvA/4cr$.

Solution 6.20

Problem 6.20

For a flat surface subject to a solar intensity, I, the pressure is given by:

$$P = \frac{F}{A} = \frac{KI}{c},$$

where K is a scattering constant with values between 0 and 2. $K = 1$ for purely absorbing; $K = 2$ for purely reflecting; $K = 0$ for transparent. So for a completely absorbing earth with $K = 1$:

$P = 4.3 \times 10^{-6}$ Pa

If earth is 38% reflecting, then K will be increased to a value > 1 but < 2. P is still very small compared with the atmospheric pressure, which is 101 kPa.

Solution 6.21

Problem 6.21

For a particle of area (normal to solar flux) A, the flux is then IA. The rate of mass change of the particle is then IA/c (kg/s) as in the previous problem ($K = 1$). We can write the equation of motion consisting of the gravitational term and the radiation pressure term as:

$$\frac{d\left(m\vec{v}\right)}{dt} = -\frac{GmM}{r^3}\vec{r} + \frac{LA}{4\pi r^3 c}\vec{r}$$

$$m\frac{d\vec{v}}{dt} = -\left(GMm - \frac{LA}{4\pi c}\right)\frac{\vec{r}}{r^3} + \frac{LA}{4\pi r^2 c^2}\vec{v}$$

Thus there will be no inward spiraling for:

$$GMm < \frac{LA}{4\pi c}$$

For a spherical particle, the cross-sectional area is simply πr^2. Expressing the mass in terms of the density gives a critical radius of the particle as:

$$r_{critical} = \frac{3}{16\pi}\frac{L}{GM\rho c}$$

For reasonable solar luminance values and orbits comparable to that of the Earth, the critical radius is about 1 micrometer. It can be seen from the equation, however, that for low-density particles, the critical radius is higher. This might imply that the value would become measurable for low-density satellites such as balloons.

Solution 6.22

Want to make use of the voltage divider formula:

$V_{out} = \dfrac{R_2}{R_1 + R_2} V_{in}$ and collapse the resistances into equivalent resistances at each point.

This gives

$$V_{out} = \frac{R}{R + R} V_2 = \frac{V_2}{2}$$

$$V_2 = \frac{R_2'}{R_1' + R_2'} V_1 = \frac{\frac{2R}{3}}{R + \frac{2R}{3}} V_1 = \frac{2}{5} V_1$$

where R_1' and R_2' are calculated by evaluating the equivalent circuit at the point V_2.
Similarly, collapse the circuit again to calculate the equivalent resistances at the point V_1:

$$V_1 = \frac{R_2''}{R_1'' + R_2''} V_{in}$$

$$\frac{1}{R_2''} = \frac{1}{R} + \frac{1}{R + R_2'} = \frac{8}{5R}$$

$$\frac{\frac{5R}{8}}{R + \frac{5R}{8}} V_1 = \frac{5}{15} V_{in}$$

$$V_{out} = \frac{1}{2} \frac{2}{5} \frac{5}{13} V_{in} = \frac{V_{in}}{13}.$$

Solution 6.23

$$\vec{B}_{out} = \vec{B}_a + \mu_0 \nabla \left(\frac{A\cos\theta}{r^2} \right)$$

To determine A, set the magnetic scalar potential:

$V_M = -B_a z + C$ at large r, $V_M = $ constant (call it 0) at $r = a$

$$\Rightarrow V_M = -B_a r\cos\theta - \frac{\mu_0 A\cos\theta}{r^2}.$$

Now require:

$$B_a r\cos\theta = \frac{\mu_0 A\cos\theta}{a^2}$$

$$A = \frac{-B_a a^3}{\mu_0}$$

and calculate:

$$\nabla\left(\frac{A\cos\theta}{r^2}\right) = A\left[\frac{-2\cos\theta}{r^3}\,\hat{r} + \frac{1}{r}\left(\frac{-\sin\theta}{r^2}\right)\hat{\theta}\right]$$

$$\Rightarrow \vec{B} = \left(B_a - \frac{2\mu_0 A}{r^3}\right)\cos\theta\,\hat{r} - \left(B_a + \frac{\mu_0 A}{r^3}\right)\sin\theta\,\hat{\theta}$$

$$= B_a\left[\left(1 + \frac{2a^3}{r^3}\right)\cos\theta\,\hat{r} + \left(1 - \frac{a^3}{r^3}\right)\sin\theta\,\hat{\theta}\right]$$

$$B(a,0) = 3Ba$$

Solution 6.24

Problem 6.24

$$\Phi = -\int k\frac{dq}{r} = \frac{-kq}{\sqrt{x^2 + r^2}}, \text{ where } k = \frac{1}{4\pi\varepsilon_0}. \text{ Then}$$

$$U = e\Phi = \frac{-keq}{\sqrt{x^2 + r^2}}\text{initially}$$

$$= \frac{-keq}{r}\text{finally}$$

$$\Delta U = \frac{1}{2}mv^2 = keq\left[\frac{1}{r} - \frac{1}{\sqrt{x^2 + r^2}}\right]$$

$$v = \sqrt{\frac{2keq}{m}}\left[\frac{1}{r} - \frac{1}{\sqrt{x^2 + r^2}}\right]^{1/2}$$

The problem is nonrelativistic as long as $v \ll c$. For large x, ignore the second term in v to give the condition:

$$\frac{v}{c} \approx \sqrt{\frac{2keq}{mrc^2}} \ll 1$$

$$\frac{q}{r} \ll \frac{mc^2}{2ke}$$

Solution 6.25

Problem 6.25

Using the "method of images," an image charge of charge $-q$ will be present a distance d below the surface of the plane. Thus the attractive force will be $F = \frac{kq^2}{(2d)^2}$. The induced charge is given by:

$$\sigma = -\varepsilon_0 \frac{\partial V}{\partial z}\bigg|_{z=0}$$

$$V = \frac{kq}{\sqrt{r^2 + (z-d)^2}} - \frac{kq}{\sqrt{r^2 + (z+d)^2}}$$

$$\Rightarrow \sigma = -\frac{1}{2\pi} \frac{dq}{(r^2 + d^2)^{3/2}}.$$

Solution 6.26

Problem 6.26

a. $\int_S \vec{D} \cdot \hat{n} \, dS = \int_V \rho \, dV$

$\int_C \vec{E} \cdot d\vec{l} = -\frac{\partial}{\partial t} \int_S \vec{B} \cdot \hat{n} \, dS$

$\int_S \vec{B} \cdot \hat{n} \, dS = 0$

$\int_C \vec{H} \cdot d\vec{l} = \int_S \vec{J} \cdot \hat{n} \, dS + \frac{\partial}{\partial t} \int_S \vec{D} \cdot \hat{n} \, dS$

b. $\int_C \vec{H} \cdot d\vec{l} = \int_S \vec{J} \cdot \hat{n} \, dS + \frac{\partial}{\partial t} \int_S \vec{D} \cdot \hat{n} \, dS$

$\vec{\nabla} \times \vec{H} = \vec{J} + \frac{\partial \vec{D}}{\partial t}$

$\int_S \vec{\nabla} \times \vec{H} \cdot \hat{n} \, dS = \int_S \vec{J} \cdot \hat{n} \, dS = \int_S \vec{J} \cdot \hat{n} \, dS + \frac{\partial}{\partial t} \int_S \vec{D} \cdot \hat{n} \, dS$

Stokes' theorem: $\int_S \vec{\nabla} \times \vec{H} \cdot \hat{n} \, dS = \int_C \vec{H} \cdot d\vec{l}$

$\int_C \vec{H} \cdot d\vec{l} = \int_S \vec{J} \cdot \hat{n} \, dS + \frac{\partial}{\partial t} \int_S \vec{D} \cdot \hat{n} \, dS$

c. Consider a parallel-plate capacitor in the center of a very large, alternating-current wire, as shown in the picture. It is easy to see that even neglecting the displacement current there will be some magnetic field on this symmetrically-placed surface from the current-carrying wires. However, neglecting the displacement current from Maxwell's laws, one would be led to believe that there is no magnetic field at all along this loop!

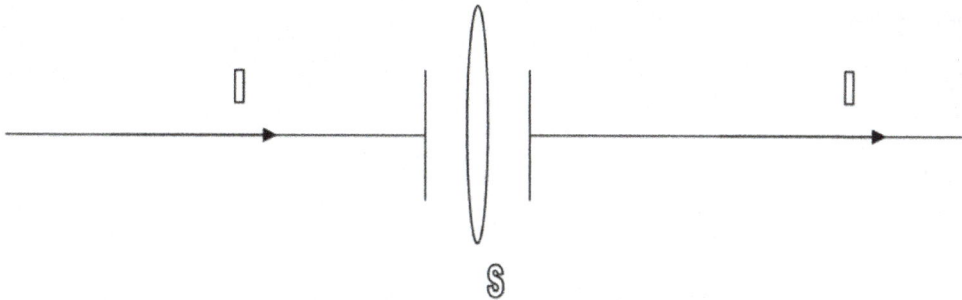

d. i. Charge and current free:

$$\vec{\nabla} \cdot \vec{D} = 0, \ \vec{\nabla} \times \vec{E} + \frac{\partial \vec{B}}{\partial t} = 0, \ \vec{\nabla} \cdot \vec{B} = 0, \ \vec{\nabla} \times \vec{H} = \frac{\partial \vec{D}}{\partial t}$$

$$\vec{D} = \varepsilon \vec{E}, \ \vec{H} = \frac{1}{\mu} \vec{B}$$

Ampere: $\dfrac{1}{\mu\varepsilon} \vec{\nabla} \times \vec{B} = \dfrac{\partial \vec{E}}{\partial t} \to \dfrac{1}{\mu\varepsilon} \vec{\nabla} \times \dfrac{\partial \vec{B}}{\partial t} \dfrac{\partial^2 \vec{E}}{\partial t^2}$

Faraday: $-\dfrac{1}{\mu\varepsilon} \vec{\nabla} \times \left(\vec{\nabla} \times \vec{E} \right) = \dfrac{\partial^2 \vec{E}}{\partial t^2} \to -\dfrac{1}{\mu\varepsilon} \left[\vec{\nabla} \left(\vec{\nabla} \cdot \vec{E} \right) - \vec{\nabla}^2 \vec{E} \right] = \dfrac{\partial^2 \vec{E}}{\partial t^2}$

Gauss: $\dfrac{1}{\mu\varepsilon} \vec{\nabla}^2 \vec{E} = \dfrac{\partial^2 \vec{E}}{\partial t^2}$

Ampere: $\dfrac{1}{\mu\varepsilon} \vec{\nabla} \times \vec{B} = \dfrac{\partial E}{\partial t} \to \dfrac{1}{\mu\varepsilon} \vec{\nabla} \times \left(\vec{\nabla} \cdot \vec{B} \right) = \dfrac{\partial}{\partial t} \left(\vec{\nabla} \cdot \vec{E} \right)$

Faraday: $\dfrac{1}{\mu\varepsilon} \left[\vec{\nabla} \left(\vec{\nabla} \cdot \vec{B} \right) - \vec{\nabla}^2 \vec{B} \right] = -\dfrac{\partial^2}{\partial t^2} \vec{B}$

Gauss/magnetism: $\dfrac{1}{\mu\varepsilon} \vec{\nabla}^2 \vec{B} = \dfrac{\partial^2}{\partial t^2} \vec{B}$

ii. $\vec{E} = \vec{E}_0 e^{i(kz - \omega t)}$

$\vec{B} = \vec{B}_0 e^{i(kz - \omega t)}$

To simplify this a bit further, take $\vec{E}_0 = E_0 \hat{x}$ and then from Faraday's law:

$\vec{\nabla} \times \vec{E} = -ikE_0 \hat{y} e^{i(kz - \omega t)}$

$\frac{\partial}{\partial t} \vec{B} = -i\omega \vec{B}_0 e^{i(kz - \omega t)}$

Faraday: $-ikE_0\hat{y} = -\left(-i\omega\vec{B}_0\right) \rightarrow \vec{B}_0 = -\dfrac{k}{\omega}E_0\hat{y}$

Also have Ampere's law:

$$\frac{1}{\mu\varepsilon}\vec{\nabla}\times\vec{B} = \frac{\partial\vec{E}}{\partial t}$$

$$\frac{1}{\mu\varepsilon}\left(-ikB_0\hat{x}\right) = i\omega E_0\hat{x}$$

$$\frac{1}{\mu\varepsilon} = \frac{\omega^2}{k^2}$$

So we are free to pick an amplitude of electric field and a wavelength or frequency, and in terms of the free variables:

$$\vec{E} = E_0\hat{x}e^{i\left(kz-\frac{k}{\sqrt{\mu\varepsilon}}t\right)}$$

$$\vec{B} = -\frac{1}{\sqrt{\mu\varepsilon}}E_0\hat{y}e^{i\left(kz-\frac{k}{\sqrt{\mu\varepsilon}}t\right)}$$

iii. The classical wave equation is:

$\vec{\nabla}^2\vec{F} = \dfrac{1}{v^2}\dfrac{\partial^2}{\partial t^2}\vec{F}$. From the form of the differential equations in part i, this indicates that

$$v = \frac{1}{\sqrt{\mu\varepsilon}}.$$

e. In a charge-free and current-free region, have:

$$\int_S \vec{D}\cdot\hat{n}dS = 0, \quad \int_C \vec{E}\cdot d\vec{l} = -\frac{\partial}{\partial t}\int_S \vec{B}\cdot\hat{n}dS,$$

$$\int_S \vec{B}\cdot\hat{n}dS = 0, \quad \int_C \vec{H}\cdot d\vec{l} = \frac{\partial}{\partial t}\int_S \vec{D}\cdot\hat{n}dS.$$

And from Gauss's laws, must have $\vec{D}_\perp^1 = \vec{D}_\perp^2$ and $\vec{B}_\perp^1 = \vec{B}_\perp^2$, since a Gaussian pillbox drawn across the surface would have to contain no net charge and no magnetic source or sink, and so, for the large faces on either side of the surface to have no net flux, these must be equal. Next, draw an Amperian loop running parallel and very close to the surface and allow it to clamp down very close to the surface. From Faraday's law, this clamps the magnetic field inside the contour down to zero, and so $\vec{E}_\parallel^1 = \vec{E}_\parallel^2$. Similarly, from Ampere's law in current-free space, $\vec{H}_\parallel^1 = \vec{H}_\parallel^2$.

f. From part d-iii, see that the speed changes. Based on part d-ii, see that since wavelength and frequency are related through the photon's speed, at least one must change. Evidently, only the wavelength changes—objects appear to be the same color underwater, despite the fact that the wave has undeniably slowed down.

g. i. Here, $\vec{J} = \sigma\vec{E}$. Further, let $\rho = 0$ and in a conductor $\vec{D} = 0$. Now

$$\vec{\nabla}\cdot\vec{E} = 0, \quad \vec{\nabla}\times\vec{E} + \frac{\partial\vec{B}}{\partial t} = 0, \quad \vec{\nabla}\cdot\vec{B} = 0, \quad \frac{1}{\mu}\vec{\nabla}\times\vec{B} = \sigma\vec{E}.$$ Starting with Ampere's law, have:

$$\frac{1}{\mu}\vec{\nabla}\times\vec{B} = \sigma\vec{E} \rightarrow \frac{1}{\mu}\vec{\nabla}\times\left(\vec{\nabla}\times\vec{E}\right) = \sigma\left(\vec{\nabla}\times\vec{E}\right)$$

Faraday: $\frac{1}{\mu}\left[\vec{\nabla}\left(\vec{\nabla}\cdot\vec{B}\right) - \vec{\nabla}^2\vec{B}\right] = -\sigma\frac{\partial\vec{B}}{\partial t}$

Gauss/magnetism: $\frac{1}{\mu}\vec{\nabla}^2\vec{B} = \sigma\frac{\partial\vec{B}}{\partial t}$

For electric field, again start with Ampere's law:

$$\frac{1}{\mu}\vec{\nabla}\times\vec{B} = \sigma\vec{E} + \varepsilon\frac{\partial\vec{E}}{\partial t} \rightarrow \frac{1}{\mu}\vec{\nabla}\times\frac{\partial\vec{B}}{\partial t} = \sigma\frac{\partial\vec{E}}{\partial t}$$

Faraday: $\frac{1}{\mu}\vec{\nabla}\times\left(-\vec{\nabla}\times\vec{E}\right) = \sigma\frac{\partial\vec{E}}{\partial t} \rightarrow -\frac{1}{\mu}\left(\vec{\nabla}\left(\vec{\nabla}\times\vec{E}\right) - \vec{\nabla}^2\vec{E}\right) = \sigma\frac{\partial\vec{E}}{\partial t}$

Gauss/electric: $\frac{1}{\mu}\vec{\nabla}^2\vec{E} = \sigma\frac{\partial\vec{E}}{\partial t}$

These differential equations have solutions of the same form and can be solved through separation of variables:

$$\frac{1}{\mu}\vec{\nabla}^2\vec{F} = \sigma\frac{\partial\vec{F}}{\partial t} = \vec{C}$$

For simplicity of solution, assign a direction to this and then generalize:

$$\frac{1}{F(x)}\left[\frac{1}{\mu}\frac{\partial^2(x)}{\partial x^2}\right] = \frac{1}{F(t)}\sigma\frac{\partial F}{\partial t} = C.$$ The solution in position x is then:

$F(x) = A_x e^{\sqrt{\mu C}x} + B_x e^{-\sqrt{\mu C}x}$. The solution in time is also a very simple damped oscillator:
$F(t) = A_t e^{\frac{C}{\sigma}t}$, and so
$E(x) \approx B(x) \approx F(x)F(t) \approx \left[A_x e^{\sqrt{\mu C}x} + B_x e^{-\sqrt{\mu C}x}\right]\left[A_t e^{\frac{C}{\sigma}t}\right]$ for some constant C. However, for this to be a propagating wave in time, the constant C must be imaginary. Thus:

$\frac{C}{\sigma}t = i\omega t$, $C = i\omega\sigma$. These constants should then be fixed in the particular system and related as in part d-ii.

ii. These fields will die off over time as the imaginary constant C will create a real term in the exponentials from part i. This real term must be negative, or otherwise it would not be physically realizable.

iii. As evidenced in the exponential from part i, the field will not completely die out, but rather exponentially die to zero.

The real part of $\sqrt{i\omega\mu\sigma}$ is $-\left|Re\left[e^{i\frac{\pi}{4}}\sqrt{\omega\mu\sigma}\right]\right| = -\sqrt{\frac{\omega\mu\sigma}{2}}$, so $\frac{1}{e}$ of the amplitude remains after

$\sqrt{\frac{2}{\omega\mu\sigma}}$ seconds. This could be regarded as analogous to the "$\frac{1}{e}$ -life" in terms of depth traveled into the conductor.

Solution 6.27

Problem 6.27

a. Radiation is emitted when the acceleration has a component in the direction of velocity. The acceleration in the rotating sphere is perpendicular to the direction that the charges are moving in, and therefore no radiation is emitted in this case.

b. First, find the magnetic moment of the sphere:

$$\vec{\mu} = \frac{1}{2}\int \vec{x} \times \vec{J}(\vec{x})d^3x$$

Thus, the magnetic moment of a loop with a uniform current I is:

$$\vec{\mu} = \frac{1}{2}\hat{n}\int (rI)rd\theta = I\pi r^2$$

Now integrating a stack of loops to obtain a sphere:

$$\vec{\mu} = \hat{n}\int_0^\pi \left[I(\phi)\pi r(\phi)^2\right]R\sin\phi d\phi$$

$$r(\phi) = R\sin\phi$$

$$I(\phi) = \frac{Q}{4\pi R^2}(\omega R\sin\phi)$$

$$\vec{\mu} = \hat{n}\frac{Q\omega R^2}{4}\int\limits_0^\pi \sin^4\phi\, d\phi = \hat{n}\frac{Q\omega R^4}{4}\left[\frac{3\pi}{8}\right]$$

Now use this to determine the torque on the sphere:

$$\tau = \left|\vec{\mu} \times \vec{B}\right| = \left|\mu\right|\left|B\right|\sin\alpha = \frac{3\pi}{32}Q\omega R^2 B\sin\alpha,\text{ where the torque will be in the direction}$$

perpendicular to both the direction of rotation and the magnetic field. Now:

$$\vec{\tau} = \frac{d\vec{L}}{dt},\ \vec{L} = I_s\vec{\omega}$$

Note that since the torque is always perpendicular to the angular momentum, the angular momentum will precess about the magnetic field and never actually change.

Finally, need to find the period of precession from this. In order to do so, consider the total amount which the angular momentum must change in one precession: since $\left|L\right| = I_s\omega$, it will trace a circle of perimeter $2\pi\left(\left|L\right|\sin\alpha\right) = 2\pi I_s\omega\sin\alpha$. Then, the time for one precession is:

$$\text{period} = \frac{2\pi I_s\omega\sin\alpha}{\tau} = \frac{64}{3\pi QR^2 B}I_s,\text{ where}$$

$$I_s = \frac{M}{\frac{4}{3}\pi R^3}\int\limits_0^R\int\limits_0^\pi\int\limits_0^{2\pi}(r\sin\phi)^2 r^2\sin\phi\, d\theta\, d\phi\, dr = \frac{MR^5}{\frac{4}{3}\pi R^3}\left(\frac{2\pi}{5}\right)\int\limits_0^\pi(\sin\phi)^2\sin\phi\, d\phi$$

$$= \frac{MR^5}{\frac{4}{3}\pi R^3}\left(\frac{2\pi}{5}\right)\left(\frac{4}{3}\right) = \frac{2}{5}MR^2,\text{ period} = \frac{128M}{15\pi QB}.$$

Note that this breaks down to division by zero if $\omega = 0$.

Solution 6.28 Problem 6.28

a. Using $\dfrac{1}{\mu_0}\oint\limits_C \vec{B}\cdot d\vec{l} = \int\limits_{S_o}\vec{J}\cdot d\vec{a} + \varepsilon_0\dfrac{\partial}{\partial t}\int\limits_{S_0}\vec{E}\cdot d\vec{a}$, define a contour to be a closed, circular

loop between the capacitor plates and centered on the z-axis, oriented such that the positive z-axis is the surface normal.

Now have:

$$\frac{1}{\mu_0}\oint_C \vec{B}\cdot d\vec{l} = \int_{S_o} \vec{J}\cdot d\vec{a} + \varepsilon_0 \frac{\partial}{\partial t}\int_{S_0} \vec{E}\cdot d\vec{a}$$

$$E(\rho,\phi,z) = E_0 f_E\left(\frac{\omega\rho}{c}\right)\cos(\omega t)\hat{k}$$

$$B(\rho,\phi,z) = -\frac{E_0}{c}f_B\left(\frac{\omega\rho}{c}\right)\sin(\omega t)\hat{e}_\phi$$

$$\vec{J} = 0$$

For:

$$\frac{1}{\mu_0}(2\pi\rho)\left(-\frac{E_0}{c}f_B\left(\frac{\omega\rho}{c}\right)\sin(\omega t)\right) = \varepsilon_0\frac{\partial}{\partial t}\int_0^\rho 2\pi\rho'\left[E_0 f_E\left(\frac{\omega\rho'}{c}\right)\cos(\omega t)\right]d\rho'$$

Simplifying:

$$\rho\left(f_B\left(\frac{\omega\rho}{c}\right)\right) = \int_0^\rho \frac{\omega\rho'}{c}\left[f_E\left(\frac{\omega\rho'}{c}\right)\right]d\rho'$$

Let:

$$\rho\left(f_E\left(\frac{\omega\rho}{c}\right)\right) = \int_0^\rho \frac{\omega\rho'}{c}\left[f_E\left(\frac{\omega\rho'}{c}\right)\right]d\rho'$$

$$x' = \frac{\omega\rho'}{c},\ dx' = \frac{\omega}{c}dp',\ x = \frac{\omega\rho}{c},\ \rho = \frac{cx}{\omega}\ (*)$$

$$xf_B(x) = \int_0^x x'\left[f_E(x')\right]dx'$$

Now use:

$$\oint_C \vec{E}\cdot d\vec{l} = -\frac{\partial}{\partial t}\int_{S_0} \vec{B}\cdot d\vec{a},$$ where we define a contour to be a rectangular region oriented

vertically between plates, with one end along the z-axis and the top edges of the contour

on the capacitor plates. Let this contour be oriented so that its surface normal is in the $\hat{\phi}$

direction. Now have:

$$dE(0) - dE(\rho) = -\frac{\partial}{\partial t}d\int_0^\rho B(\rho')d\rho'$$

$$E(\rho, \phi, z) = E_0 f_E\left(\frac{\omega\rho}{c}\right)\cos(\omega t)\hat{k}$$

$$B(\rho, \phi, z) = -\frac{E_0}{c} f_B \frac{\omega\rho}{c}\sin(\omega t)\hat{e}_\phi$$

$$f_E(0) = 1$$

$$f_E(0)\cos(\omega t) - f_E\left(\frac{\omega\rho}{c}\right)\cos(\omega t) = \frac{\partial}{\partial t}\frac{1}{c}\int_0^\rho f_B\left(\frac{\omega\rho'}{c}\right)\sin(\omega t)d\rho'$$

$$1 - f_E\left(\frac{\omega\rho}{c}\right) = \frac{\omega}{c}\int_0^\rho f_B\left(\frac{\omega\rho'}{c}\right)d\rho'$$

Again using (*), we obtain

$$1 - f_E(x) = \int_0^x f_B(x)dx', \text{ as expected.}$$

Finally, take:

$$f_E(x) = 1 - \int_0^x dx' f_B(x')$$

$$x f_B(x) = \int_0^x dx' x' f_E(x')$$

For:

$$f_E'(x) = -f_B(x)$$
$$f_E''(x) = -f_B'(x)$$
$$f_B(x) + x f_B'(x) = x f_E(x)$$
$$2f_B'(x) + x f_B''(x) = f_E(x) + x f_E'(x)$$

Combining:

$$\left[-f_E'(x)\right] + x\left[-f_E''(x)\right] = x f_E(x)$$
$$x f_E(x) + f_E'(x) + x f_E''(x) = 0$$
$$x^2 f_E(x) + x f_E'(x) + x^2 f_E''(x) = 0$$
$$f_E(x) \to J_0(x)$$

and

$$2f_B'(x) + x f_B''(x) = \frac{1}{x}\left(f_B(x) + x f_B'(x)\right) + x\left(-f_B(x)\right)$$

$$2x f_B'(x) + x^2 f_B''(x) = f_B(x) + x f_B'(x) - x^2 f_B(x')$$

$$x^2 f_B(x) - f_B(x) + x f_B'(x) + x^2 f_B''(x) = 0$$

$$(x^2 - 1) f_B'(x) + x^2 f_B''(x) = 0$$

$$f_B(x) \to J_1(x)$$

b. Consider:

$\varepsilon_0 \int_{S_c} \overrightarrow{E} \cdot d\overrightarrow{a} = \int_V \rho dV$. Ignore all fringe effects and define a pillbox-shaped surface to

encompass one of the two capacitor plates, with the circular ends parallel to the capacitor plate. Now, it is easy to see that:

$$I(t) = I_0 \sin(\omega t) = \frac{\partial \rho_{\text{total, 1 plate}}(t)}{\partial t}$$

$$\rho_{\text{total, 1 plate}}(t) = -\frac{I_0}{\omega}\cos(\omega t)$$

$$\varepsilon_0 \int_{S_c} \overrightarrow{E} \cdot d\overrightarrow{a} = \int_V \rho dV = \rho_{\text{total, 1 plate}}(t) = -\frac{I_0}{\omega}\cos(\omega t)$$

$$= \varepsilon_0 \int_0^{2\pi} \int_0^{\rho} \left[E_0 f_E\left(\frac{\omega \rho'}{c}\right) \cos(\omega t) \hat{k} \cdot \left(-\hat{k}\right) \right] \rho' d\rho' d\theta = -2\pi\varepsilon_0 E_0 \cos(\omega t) \int_0^a J_0\left(\frac{\omega \rho'}{c}\right) \rho' d\rho'$$

Substitute $u = \dfrac{\omega \rho'}{c}$, $du = \dfrac{\omega \rho'}{c} d\rho'$

$$= -2\pi\varepsilon_0 E_0 \cos(\omega t) \frac{c^2}{\omega^2} \int_{\rho'=0}^{\rho'=a} J_0(u) u \, du$$

$$= -2\pi\varepsilon_0 E_0 \cos(\omega t) \frac{c^2}{\omega^2} \left(J_1\left(\frac{\omega a}{c}\right)\frac{\omega a}{c} \right) = -2\pi\varepsilon_0 E_0 \cos(\omega t) \frac{ca}{\omega} \left(J_1\left(\frac{\omega a}{c}\right) \right)$$

Combining this with the earlier portion from the integral charge density, have:

$$-\frac{I_0}{\omega}\cos(\omega t) = -2\pi\varepsilon_0 E_0 \cos(\omega t) \frac{ca}{\omega}\left(J_1\left(\frac{\omega a}{c}\right) \right)$$

$$I_0 = 2\pi\varepsilon_0 E_0 ca \left(J_1\left(\frac{\omega a}{c}\right) \right), \text{ or } \frac{I_0}{2\pi\varepsilon_0 ca J_1\left(\frac{\omega a}{c}\right)} = E_0 \text{ as expected.}$$

c. $E(\rho,\phi,z) = E_0 J_0\left(\dfrac{\omega\rho}{c}\right)\cos(\omega t)\hat{k}$

$B(\rho,\phi,z) = -\dfrac{E_0}{c}J_1\left(\dfrac{\omega\rho}{c}\right)\sin(\omega t)\hat{e}_\phi$

Now, integrating to find the energy of the electric and magnetic fields in the capacitor:

$$\int E^2(\rho,\phi,z)d^3x \approx E_0^2\cos^2(\omega t)(2\pi d)\int_0^a\left(1 - \frac{1}{4}\left(\frac{\omega\rho}{c}\right)^2\right)^2\rho\, d\rho$$

$$= E_0^2\cos^2(\omega t)(2\pi d)\int_0^a\left(1 - \frac{1}{2}\left(\frac{\omega\rho}{c}\right)^2 + \frac{1}{16}\left(\frac{\omega\rho}{c}\right)^4\right)\rho\, d\rho$$

$$= E_0^2\cos^2(\omega t)(2\pi d)\left[\frac{a^2}{2} - \frac{1}{2}\left(\frac{\omega}{c}\right)^2\frac{a^4}{4} + \frac{1}{16}\left(\frac{\omega}{c}\right)^4\frac{a^6}{6}\right]$$

And for the magnetic field:

$$\int B^2(\rho,\phi,z)d^3x \approx \left(\frac{E_0}{c}\right)^2\sin^2(\omega t)(2\pi d)\int_0^a\left(\frac{1}{2}\frac{\omega\rho}{c} - \frac{1}{16}\left(\frac{\omega\rho}{c}\right)^3\right)^2\rho\, d\rho$$

$$= \left(\frac{E_0}{c}\right)^2\sin^2(\omega t)(2\pi d)\int_0^a\left(\frac{1}{4}\left(\frac{\omega\rho}{c}\right)^2 - \frac{1}{16}\left(\frac{\omega\rho}{c}\right)^4 + \frac{1}{256}\left(\frac{\omega\rho}{c}\right)^6\right)\rho\, d\rho$$

$$= \left(\frac{E_0}{c}\right)^2\sin^2(\omega t)(2\pi d)\left[\frac{1}{4}\left(\frac{\omega}{c}\right)^2\frac{a^4}{4} - \frac{1}{16}\left(\frac{\omega}{c}\right)^4\frac{a^6}{6} + \frac{1}{256}\left(\frac{\omega}{c}\right)^6\frac{a^8}{8}\right]$$

Now, using the formulae for energy density:

$$\frac{1}{2\mu_0}\int \vec{B}^2 d^3x = \frac{1}{2}LI^2$$

$$\left(\frac{E_0}{c}\right)^2\sin^2(\omega t)(2\pi d)\left[\frac{1}{4}\left(\frac{\omega}{c}\right)^2\frac{a^4}{4} - \frac{1}{16}\left(\frac{\omega}{c}\right)^4\frac{a^6}{6} + \frac{1}{256}\left(\frac{\omega}{c}\right)^6\frac{a^8}{8}\right] = \frac{1}{2}LI_0^2\sin^2(\omega t)$$

$$2\left(\frac{E_0}{I_0 c}\right)^2(2\pi d)\left[\frac{1}{4}\left(\frac{\omega}{c}\right)^2\frac{a^4}{4} - \frac{1}{16}\left(\frac{\omega}{c}\right)^4\frac{a^6}{6} + \frac{1}{256}\left(\frac{\omega}{c}\right)^6\frac{a^8}{8}\right] = L$$

Taking leading order in $\frac{\omega a}{c}$, have

$$\frac{\pi d}{4\omega^2}\left(\frac{E_0}{I_0}\right)^2\left(\frac{\omega a}{c}\right)^4 = L$$ where E_0 is defined above. Further,

$$\frac{1}{2}\varepsilon_0\int \vec{E}^2 d^2x = \frac{1}{2}\frac{Q^2}{C}$$

$$Q = \int I(t) = -\frac{I_0}{\omega}\cos(\omega t)$$

$$\frac{1}{2}\varepsilon_0 E_0{}^2\cos^2(\omega t)(2\pi d)\left[\frac{a^2}{2} - \frac{1}{2}\left(\frac{\omega}{c}\right)^2\frac{a^4}{4} + \frac{1}{16}\left(\frac{\omega}{c}\right)^4\frac{a^6}{6}\right] = \frac{1}{2C}\left(\frac{I_0}{\omega}\right)^2\cos^2(\omega t)$$

$$\varepsilon_0\left(\frac{\omega E_0}{I_0}\right)^2(2\pi d)\left[\frac{a^2}{2} - \frac{1}{2}\left(\frac{\omega}{c}\right)^2\frac{a^4}{4} + \frac{1}{16}\left(\frac{\omega}{c}\right)^4\frac{a^6}{6}\right] = \frac{1}{C}$$

To leading order:

$$\varepsilon_0 c^2\left(\frac{E_0}{I_0}\right)^2(2\pi d)\left(\frac{\omega a}{c}\right)^2 = \frac{1}{C}.$$

Now consider the natural frequency of an *LC* circuit:

$$\frac{1}{C}Q(t) + LQ''(t) = 0$$

$$-\frac{1}{LC}Q(t) = Q''(t)$$

So that solutions are clearly of the form: $Q(t) = A\sin\left(\frac{1}{\sqrt{LC}}t\right) + B\cos\left(\frac{1}{\sqrt{LC}}t\right)$. The

frequency in cycles per second is $\frac{1}{2\pi}\frac{1}{\sqrt{LC}}$. Then have:

$$\frac{1}{2\pi}\sqrt{\frac{\varepsilon_0 c^2\left(\frac{E_0}{I_0}\right)^2(2\pi d)\left(\frac{\omega a}{c}\right)^2}{\frac{\pi d}{4\omega^2}\left(\frac{E_0}{I_0}\right)^2\left(\frac{\omega a}{c}\right)^4}} = \frac{c}{2\pi}\sqrt{8\varepsilon_0}\left(\frac{c}{a}\right) = \frac{1}{\pi}\sqrt{\frac{2}{\mu_0}}\left(\frac{c}{a}\right).$$

d. Neglecting time-retardation effects, have $\vec{p} = q\vec{d}$. Taking $\rho_{\text{total,1 plate}}(t) = -\frac{I_0}{\omega}\cos(\omega t)$, get

$$\vec{p}_0 = -\vec{d}\frac{I_0}{\omega}.$$

e. Using integration by parts:

$$\int d^3r \vec{J}(\vec{r})$$

$$u = \vec{J}\left(\vec{r}\right), \, dv = d^3r, \, du = \vec{\nabla}\cdot\vec{J}\left(\vec{r}\right), \, v = \vec{r}\,d^3r$$

$$= \vec{r}\vec{J}(\vec{r})\Big|_{-\infty}^{\infty} - \int \vec{r}\left(\vec{\nabla}\cdot\vec{J}(\vec{r})\right)d^3r = -\int \vec{r}\left(\vec{\nabla}\cdot\vec{J}(\vec{r})\right)d^3r$$

$$iw\,\vec{\rho} = \left(\vec{\nabla}\cdot\vec{J}\right) \text{ (continuity equation)}$$

$$= -i\omega \int \vec{r}\rho(\vec{r})d^3r \text{ (definition of dipole moment)}$$

$$= -i\omega\vec{p}_0 \text{ which, upon substitution, gives the expected result.}$$

f. Taking $\vec{\nabla}\times\vec{A}$, observe that only terms related to $\frac{\partial}{\partial r}$ will survive. Since the vector portion $\vec{p}_0$ is not a function of r, it is possible to pull out the part that is related to r and allow the cross product to act between the r unit vector, which certainly may not survive this cross product since again r is the only variable appearing in the vector potential:

$$\vec{B}(\vec{r}) = -\frac{i\omega}{4\pi\varepsilon_0 c^2}\frac{\partial}{\partial t}\left(\frac{e^{\frac{i\omega r}{c}}}{i}\right)\left(\hat{e}_r\times\vec{p}_0\right) = \frac{i\omega}{4\pi\varepsilon_0 c^2}\frac{\partial}{\partial t}\left(\frac{e^{\frac{i\omega r}{c}}}{r}\right)\vec{p}_0\times\hat{e}_r$$

The radiation field is the portion of this that falls off as r^{-1}:

$$\vec{B}_{rad}(\vec{r}) = \frac{i\omega}{4\pi\varepsilon_0 c^2}\left(\frac{i\omega}{c}\right)\left(\frac{e^{\frac{i\omega r}{c}}}{r}\right)\vec{p}_0\times\hat{e}_r = -\frac{\omega^2\left(\vec{p}_0\times\hat{e}_r\right)}{4\pi\varepsilon_0 c^3}\left(\frac{e^{\frac{i\omega r}{c}}}{r}\right)$$

Now, the Poynting vector has units in watts per square meter. By integrating over the surface $\frac{1}{\text{cycle}}\int_{\text{cycle}}\vec{S}\cdot d\vec{S}$, should be able to get power radiated.

Certainly, from Maxwell's equations, $\frac{1}{c}\frac{\partial}{\partial t}\vec{E} = \vec{\nabla}\times\vec{B}$ in the far zone where there is no stray charge or current. Based on this, have:

$$\frac{-i\omega}{c}\vec{E} = \vec{\nabla}\times\vec{B}$$

$$\vec{E}_{rad} = -\left(\frac{1}{4\pi\varepsilon_0 c}\right)\left(\frac{i\omega}{c}\right)^2\left(\frac{e^{\frac{i\omega r}{c}}}{r}\right)\hat{e}_r\times\left(\vec{p}_0\times\hat{e}_r\right)$$

Now:

$$\vec{S} = \frac{1}{\mu_0}\left[-\left(\frac{1}{4\pi\varepsilon_0 c}\right)\left(\frac{i\omega}{c}\right)^2\left(\frac{e^{\frac{i\omega r}{c}}}{r}\right)\cdot\left(-\frac{\omega^2}{4\pi\varepsilon_0 c^3}\left(\frac{e^{\frac{i\omega r}{c}}}{r}\right)\right)\right]\left[\left(\hat{e}_r\times\left(\vec{p}_0\times\hat{e}_r\right)\right)\times\left(\vec{p}_0\times\hat{e}_r\right)\right]$$

$$\vec{S}(t) = \frac{1}{4\pi\varepsilon_0}\left[-\frac{\omega^4}{4\pi c^4}\right]\frac{e^{2\frac{i\omega r}{c}}}{r^2}e^{-2i\omega t}\left[\left(\hat{e}_r\times\left(\vec{p}_0\times\hat{e}_r\right)\right)\times\left(\vec{p}_0\times\hat{e}_r\right)\right]$$

Simplifying this with the triple cross product formula:

$$\vec{A}\times\left(\vec{B}\cdot\vec{C}\right) = \left(\vec{A}\cdot\vec{C}\right)\vec{B} - \left(\vec{A}\cdot\vec{B}\right)\vec{C},$$

$$\vec{S}(t) = \frac{1}{4\pi\varepsilon_0}\left[-\frac{\omega^4}{4\pi c^4}\right]\frac{e^{2\frac{i\omega r}{c}}}{r^2}e^{-2i\omega t}\left[\vec{p}_0 - \left(\vec{p}_0\cdot\hat{e}_r\right)\hat{e}_r\times\left(\vec{p}_0\times\hat{e}_r\right)\right]$$

$$= \frac{1}{4\pi\varepsilon_0}\left[-\frac{\omega^4}{4\pi c^4}\right]\frac{e^{2\frac{i\omega r}{c}}}{r^2}e^{-2i\omega t}\left[\left(\vec{p}_0\cdot\hat{e}_r\right)\vec{p}_0 - \left(\vec{p}_0\cdot\hat{e}_r\right)\vec{p}_0 + \left(\vec{p}_0\cdot\hat{e}_r\right)^2\hat{e}_r\right]$$

$$= \frac{1}{4\pi\varepsilon_0}\left[-\frac{\omega^4}{4\pi c^4}\right]\frac{e^{2\frac{i\omega r}{c}}}{r^2}e^{-2i\omega t}\left(\vec{p}_0\cdot\hat{e}_r\right)^2\hat{e}_r$$

Letting the capacitor very sinusoidally, the integral over one cycle time-wise becomes:

$$\int_0^{\frac{2\pi}{\omega}}\sin^2(\omega t)dt\int_0^{\frac{2\pi}{\omega}}\frac{1-\cos(2\omega t)}{2}dt = \frac{1}{2}\left(\frac{2\pi}{\omega}\right) - \frac{1}{2}\sin(4\pi) = \frac{\pi}{\omega}$$

Performing the spatial integration:

$$P = \int\frac{1}{4\pi\varepsilon_0}\left[-\frac{\omega^4}{4\pi c^4}\right]\frac{e^{2\frac{i\omega r}{c}}}{r^2}\left(-\frac{\pi}{\omega}\right)\left(\vec{p}_0\cdot\hat{e}_r\right)^2\hat{e}_r r^2\sin\theta\cdot d\vec{S}$$

It is possible to cancel and make some substitutions, fixing the sphere's surface at a particular radius where $\frac{2\omega r}{c} = 2\pi k$ (proposing one observes this as the sphere by which flux passes).

$$P = \int\frac{1}{4\pi\varepsilon_0}\left[\frac{\omega^3}{4c^4}\right]\left|\vec{p}_0\right|^2\cos^2\theta\sin\theta\cdot d\vec{S}$$

$$\int d\phi = 2\pi, \int_0^{\pi}\cos^2\theta\sin\theta d\theta = -\frac{\cos^3\pi}{3} + \frac{\cos^3 0}{3} = \frac{2}{3}, \text{ giving at last}$$

$$\langle P\rangle = \frac{\omega^3}{12c^4}\left|\vec{p}_0\right|^2.$$

Solution 6.29

Problem 6.29

a. $Q = \int \rho d^3x = \int \frac{1}{r} x \delta(x-a) d^3x = 0$

$\vec{P}_x = \int \frac{1}{r} x^2 \delta(r-a) d^3x = \sigma_0 a^3 \int \sin^3\theta \cos^2\phi \, d\theta d\phi = \frac{4}{3}\sigma_0 a^3$

The terms P_y, P_z vanish since they are integrals over odd functions. Similarly, the quadrupole terms must all die since the delta term will always contain an odd number of x factors, and the first term must always contain an odd number of some term and so the zero-centered integral over it will always disappear.

b. $\vec{B}_{avg} = \frac{3}{4\pi a^3} \int \left(\vec{\nabla} \times \vec{A} \right) d^3x = -\frac{3}{4\pi a^3} \oint \vec{A} \times d\vec{s} = -\frac{3}{4\pi a^3} \oint \frac{\mu_0}{4\pi} \int \frac{\vec{J}\left(\vec{x}',t\right)}{\left| \vec{x} - \vec{x}' \right|} d^3x' \times \hat{r} a^2 d\Omega$

Since $\int \frac{\hat{r} d\Omega}{\left| \vec{x} - \vec{x}' \right|} = \frac{4\pi}{3a^2} x'$,

$\vec{B}_{avg} = \frac{-3}{4\pi a^3} \frac{\mu_0}{4\pi} \frac{4\pi}{3a^2} \left[-2a^2 \int \frac{1}{2}\left(\vec{x}' \times \vec{J}\left(\vec{x}',t\right) \right) d^3\vec{x}' \right] = \frac{\mu_0}{4\pi} \frac{2\vec{m}}{a^3}.$

Solution 6.30

Problem 6.30

The relevant Maxwell's equations are

$\frac{1}{\mu_0} \vec{\nabla} \times \vec{B} - \frac{\partial \vec{D}}{\partial t} = 0$

$\vec{\nabla} \times \vec{E} - \frac{\partial \vec{B}}{\partial t} = 0$

Taking del-cross the second and substituting the first gives

$\vec{\nabla} \times \left(\vec{\nabla} \times \vec{E} \right) + \frac{\partial}{\partial t}\left(\vec{\nabla} \times \vec{B} \right) = 0$

$\vec{\nabla}\left(\vec{\nabla} \cdot \vec{E} \right) - \vec{\nabla}^2 \vec{E} + \mu_0 \frac{\partial}{\partial t}\left(\frac{\partial \vec{D}}{\partial t} \right) = 0$

$$\vec{\nabla}\left(\vec{\nabla}\cdot\vec{E}\right) - \vec{\nabla}^2\vec{E} + \mu_0\frac{\partial^2}{\partial t^2}\left(\varepsilon_0\vec{E} + i\vec{E}\times\vec{g}\right) = 0$$

Now we need to impose the simplifying constraints. Choose k and g to be in the z direction, and so $\vec{E}_0 = a\hat{x} + b\hat{y}$. Now we have that

$$\vec{\nabla}\left(\vec{\nabla}\cdot\vec{E}\right) = \vec{\nabla}\left(\vec{\nabla}\cdot\vec{E}_0 e^{i(kz-\omega t)}\right) = 0$$

$$\vec{\nabla}^2\vec{E} = -\vec{E}_0 k^2 e^{i(kz-\omega t)}$$

$$\frac{\partial^2}{\partial t^2}\vec{E} = -\vec{E}_0\omega^2 e^{i(kz-\omega t)}$$

giving

$-k^2\vec{E} + \mu_0\omega^2\left(\varepsilon_0\vec{E} + i\vec{E}\times\vec{g}\right) = 0$. Substituting the formula for E_0 gives a homogeneous system of equations:

$$-k^2\left(a\hat{x} + b\hat{y}\right) + \mu_0\omega^2\left[\varepsilon_0\left(a\hat{x} + b\hat{y}\right) + i\left(a\hat{x} + b\hat{y}\right)\times g\hat{z}\right] = 0$$

$$-k^2\left(a\hat{x} + b\hat{y}\right) + \mu_0\omega^2\left[\varepsilon_0\left(a\hat{x} + b\hat{y}\right) + ig\left(b\hat{x} - a\hat{y}\right)\right] = 0$$

$$-k^2 a + \mu_0\omega^2\varepsilon_0 a + i\mu_0\omega^2 g b = 0$$

$$-k^2 b + \mu_0\omega^2\varepsilon_0 b - i\mu_0\omega^2 g a = 0$$

The system can be solved with elimination of a and b to give:

$$-k^2 + \mu_0\omega^2\left(\varepsilon_0 + \frac{\mu_0\omega^2 g^2}{k^2 - \varepsilon_0\mu_0\omega^2}\right) = 0$$

$$k^2 = \mu_0\omega^2\left(\varepsilon_0 \pm g\right)$$

Since $n \equiv \frac{kc}{\omega}$, this imposes the constraints: $n = \pm c\sqrt{\mu_0\left(\varepsilon_0 \pm g\right)}$.

Solution 6.31

Problem 6.31

a. Note that $u \cdot u = c^2$. Now $r^\mu = x^\mu - \frac{1}{c^2}(u \cdot x)u^\mu$ is x^μ minus its projection onto the time-like u^μ space. Since the time-like element is removed, the only part left can be space-like.

b. $F^{\mu\nu} = \partial^\mu A^\nu - \partial^\nu A^\mu = \begin{bmatrix} 0 & \dfrac{-E_x}{c} & \dfrac{-E_y}{c} & \dfrac{-E_z}{c} \\ \dfrac{E_x}{c} & 0 & -B_z & B_y \\ \dfrac{E_y}{c} & B_z & 0 & -B_x \\ \dfrac{E_z}{c} & -B_y & B_x & 0 \end{bmatrix}$

$$\vec{E}_x = \partial^x A^0 - \partial^0 A^x$$

$$x^\mu = \left(ct,\ \vec{x} \right) \quad u^\mu = \lambda\left(c,\ \vec{v} \right)$$

$$\vec{E}_c = \partial_c A^0 - \partial_0 A^c$$

$$\vec{E}_c = \frac{1}{4\pi\varepsilon_0}\left(\partial_c \frac{qu^0}{4\pi\varepsilon_0 r} - \partial_0 \frac{qu^c}{4\pi\varepsilon_0 r} \right) = \frac{1}{4\pi\varepsilon_0}\left(\partial_c \frac{\lambda qc}{4\pi\varepsilon_0 r^2} - \partial_0 \frac{qu^c}{4\pi\varepsilon_0 r} \right).$$

Note that the scalar product in the denominator is invariant under Lorentz transformation. As usual, velocity is the time derivative of the position:

$$\partial_0 x^i = u^i$$

$$\partial^\mu \frac{1}{r} = \frac{\left(1 + \frac{u^0 u_0}{c^2} \right) u^\mu}{r^2}$$

$$\partial_c \frac{1}{r} = \frac{1 - \frac{u^c x_c u^\mu}{c^2}}{r^2}$$

Note the extra minus sign from the covariant dot product. Now by substituting these into the formulas for the fields, the equations become:

$$\vec{E}_c = \frac{q}{4c\pi\varepsilon_0 r^2}\left[\left(1 - \frac{u^c x_c}{c^2} \right) u^0 - \left(1 + \frac{u^0 u_0}{c^2} \right) u^c \right]$$

$$\vec{B}_c = \frac{q}{4\pi\varepsilon_0 r^2}\left[\left(1 - \frac{u^{c-1} u_{c-1}}{c^2} \right) u^{c+1} - \left(1 - \frac{u^{c+1} u_{c+1}}{c^2} \right) u^{c-1} \right]$$

c. For a particle at rest, $u=(c,0,0,0)$. Clearly both of the terms in the magnetic field vanish, since all nonzero-indexed values of u are zero. The electric field becomes:

$$\vec{E}_c = \frac{q}{4c\pi\varepsilon_0 r^2}(c) = \frac{q}{4\pi\varepsilon_0 r^2},\ \text{as expected.}$$

Solution 6.32

$$\frac{\partial}{\partial t}\rho\left(\vec{x},t\right) = \frac{\partial}{\partial t}q\delta^3\left(\vec{x}-\vec{s}(t)\right) = \frac{\partial}{\partial t}q\delta(x_x-(x_{0x}+v_xt))\delta\left(x_y-(x_{0y}+v_yt)\right)\delta(x_z-(x_{0z}+v_zt))$$

$$= q\left[\begin{array}{l} -\delta'(x_x-s_x(t))s_x'(t)\delta\left(x_y-(x_{0y}+v_yt)\right)\delta(x_z-(x_{0z}+v_zt)) - \delta'\left(x_y-s_y(t)\right)s_y'(t)\delta(x_x-(x_{0x}+v_xt))\delta(x_z-(x_{0z}+v_zt)) \\ -\delta'(x_z-s_z(t))s_z'(t)\delta\left(x_y-(x_{0y}+v_yt)\right)\delta(x_x-(x_{0x}+v_xt)) \end{array}\right]$$

$$= -q\vec{v}\left[\begin{array}{l} \delta'(x_x-s_x(t))\delta\left(x_y-(x_{0y}+v_yt)\right)\delta(x_z-(x_{0z}+v_zt)) + \delta'\left(x_y-s_y(t)\right)\delta(x_x-(x_{0x}+v_xt))\delta(x_z-(x_{0z}+v_zt)) \\ +\delta'(x_z-s_z(t))\delta\left(x_y-(x_{0y}+v_yt)\right)\delta(x_x-(x_{0x}+v_xt)) \end{array}\right]$$

$$\vec{\nabla}\cdot\vec{J} = \vec{\nabla}\cdot q\vec{v}\,\delta^3\left(\vec{x}-\vec{s}(t)\right)$$

$$= -q\vec{v}\left[\begin{array}{l} \frac{\partial}{\partial x}\delta(x_x-s_x(t))\delta\left(x_y-(x_{0y}+v_yt)\right)\delta(x_z-(x_{0z}+v_zt)) + \frac{\partial}{\partial y}\delta(x_x-(x_{0x}+v_xt))\delta\left(x_y-(x_{0y}+v_yt)\right)\delta(x_z-(x_{0z}+v_zt)) \\ +\frac{\partial}{\partial z}\delta(x_x-(x_{0x}+v_xt))\delta\left(x_y-(x_{0y}+v_yt)\right)\delta(x_z-(x_{0z}+v_zt)) \end{array}\right]$$

$$= -q\vec{v}\left[\begin{array}{l} \delta'(x_x-s_x(t))\delta\left(x_y-(x_{0y}+v_yt)\right)\delta(x_z-(x_{0z}+v_zt)) + \delta(x_x-(x_{0x}+v_xt))\delta'\left(x_y-(x_{0y}+v_yt)\right)\delta(x_z-(x_{0z}+v_zt)) \\ +\delta(x_x-(x_{0x}+v_xt))\delta\left(x_y-(x_{0y}+v_yt)\right)\delta'(x_z-(x_{0z}+v_zt)) \end{array}\right]$$

Now by examination of these two results, it is easy to see that $\vec{\nabla}\cdot\vec{J} = -\frac{\partial\rho}{\partial t}$.

Solution 6.33

To show this, make use of Green's second identity:

$$\int_V\left(\phi\nabla^2\psi-\psi\nabla^2\phi\right) = \oint_S\left(\phi\frac{\partial\psi}{\partial n'}-\psi\frac{\partial\phi}{\partial n'}\right)da'$$

Now let $\psi = \frac{1}{|x-x'|}$, and we know that $\nabla^2\phi = 0$ because the space is charge-free. Now:

$$\nabla^2\frac{1}{|x-x'|} = -4\pi\delta(x-x')$$

$$\oint_S(-\nabla\phi\cdot\hat{n})da = \oint_S\left(\vec{E}\cdot\hat{n}\right)da = 0$$

Now making use of the Green's identity gives:

$$-4\pi\phi(x) = \oint_S \left(\phi \frac{\partial}{\partial n'} \frac{1}{|x-x'|} - \frac{1}{|x-x'|} \frac{\partial\phi}{\partial n'} \right) d\theta d\phi.$$

For convenience, assume the point of interest is at the origin and the desired sphere is of radius R.

$$-4\pi\phi(x) = \oint_S \left(\phi \frac{\partial}{\partial n'} \frac{1}{r} - \frac{1}{r} \frac{\partial\phi}{\partial n'} \right) d\theta d\phi$$

$$= \oint_S \phi \left(\nabla\left(\frac{1}{r}\right) \cdot \hat{r} \right) d\theta d\phi - \oint_S \left(\frac{1}{r}\right) \frac{\partial\phi}{\partial n'} d\theta d\phi$$

$$= \oint_S \phi \left(\nabla\left(\frac{1}{r}\right) \cdot \hat{r} \right) d\theta d\phi - R\oint_S (\nabla\phi \cdot \hat{r}) d\theta d\phi$$

In the right surface integral, note that it is essentially the integral of electric field over a sphere. Since the sphere encompasses charge-free space, this integral goes to zero.

$$-4\pi\phi(x) = \oint_S \phi \left(\nabla\left(\frac{1}{r}\right) \cdot \hat{r} \right) d\theta d\phi = \oint_S \phi \left(\frac{\partial}{\partial r}\left(\frac{1}{r}\right) \hat{r} \cdot \hat{r} \right) d\theta d\phi = \frac{1}{4\pi R^2} \oint_S \phi d\theta d\phi$$

This corresponds to the average potential along the sphere.

Solution 6.34

Problem 6.34

a. The retarded vector potential has the form

$$\vec{A}(\vec{x},t) = \frac{\mu_0}{4\pi} \int \frac{\vec{J}\left(\vec{x}',t-\frac{R}{c}\right)}{R} d^3x$$

Let the wire be along the z axis, with the current running in the z direction. In this case, there is cylindrical symmetry, so the potential is a function of distance from the wire only.

$$\vec{A}(r,t) = \frac{\mu_0 \hat{z}}{4\pi} \int \frac{J\left(z,t-\frac{\sqrt{r^2+z^2}}{c}\right)}{\sqrt{r^2+z^2}} dz = \frac{\mu_0 \hat{z}}{4\pi} \int_{-\sqrt{t^2c^2-r^2}}^{\sqrt{t^2c^2-r^2}} \frac{k\left(t-\frac{\sqrt{r^2+z^2}}{c}\right)}{\sqrt{r^2+z^2}}$$

This is only valid when the limits of integration are real; otherwise the vector potential is zero.

$$\vec{A}(r,t) = \frac{\mu_0 \hat{z}}{4\pi} \int\limits_{-\sqrt{t^2c^2-r^2}}^{\sqrt{t^2c^2-r^2}} \left[\frac{kt}{\sqrt{r^2+z^2}} - \frac{k}{c}\right] dz = \frac{\mu_0 \hat{z}}{4\pi}\left[kt\ln\left(\frac{tc+\sqrt{t^2c^2-r^2}}{tc-\sqrt{t^2c^2-r^2}}\right) - \frac{2k}{c}\sqrt{t^2c^2-r^2}\right]$$

b. In this cylindrical coordinate system, have:

$$\vec{B} = \vec{\nabla}\times\vec{A} = -\widehat{\phi}\,\frac{\partial}{\partial r}\frac{\mu_0}{4\pi}\left[kt\ln\left(\frac{tc+\sqrt{t^2c^2-r^2}}{tc-\sqrt{t^2c^2-r^2}}\right) - \frac{2k}{c}\sqrt{t^2c^2-r^2}\right] = \widehat{\phi}\,\frac{\mu_0 k}{2\pi c}\left[\frac{\sqrt{t^2c^2-r^2}}{r}\right]$$

$$\vec{E} = -\frac{\partial\vec{A}}{\partial t} = -\frac{\mu_0 k\hat{z}}{4\pi}\ln\left[\left(\frac{2tc\left(tc+\sqrt{t^2c^2-r^2}\right)}{r^2}\right) - 1\right]$$

Solution 6.35

Problem 6.35

Exploit the cylindrical symmetry of the problem to write

$$\vec{A}\left(\vec{x},t\right) = \frac{\mu_0}{4\pi}\int\frac{\vec{J}\left(\vec{x},t-\frac{R}{c}\right)}{R}d^3x = \frac{\mu_0\widehat{\theta}}{4\pi}\int\limits_0^{2\pi}\frac{I_0\cos(\omega t - kR)}{R}a\cos\theta d\theta, \text{ where}$$

$R = \sqrt{z^2 + (r-a\cos\theta)^2 + (a\sin\theta)^2}$. Since $R \gg a$, expand in a to get (letting $D = \sqrt{z^2+r^2}$):

$R \approx D - \frac{ra\cos\theta}{D}$ $\frac{1}{R} \approx \frac{1}{D} + \frac{ra\cos\theta}{D^3}$. Making these substitutions, have:

$$\vec{A}\left(\vec{x},t\right) = \frac{\mu_0 I_0 a\widehat{\theta}}{4\pi}\int\limits_0^{2\pi}\cos\left(\omega t - kD + \frac{rka\cos\theta}{D}\right)\left[\frac{1}{D} + \frac{ra\cos\theta}{D^3}\right]\cos\theta d\theta$$

Now expanding in a:

$$\cos\left(\omega t - kD + \frac{rka\cos\theta}{D}\right) \approx \cos(-\omega t + kD) + \frac{kra}{D}\cos\theta\sin(-\omega t + kD)$$

This yields the expression:

$$\vec{A}\left(\vec{x},t\right) = \frac{\mu_0 I_0 a\widehat{\theta}}{4\pi}\int\limits_0^{2\pi}\left[\cos(-\omega t + kD) + \frac{kra}{D}\cos\theta\sin(-\omega t + kD)\right]\left[\frac{1}{D} + \frac{ra\cos\theta}{D^3}\right]\cos\theta d\theta$$

Integrating:

$$\vec{A}\left(\vec{x},t\right) = \frac{\mu_0 I_0 a \hat{\theta}}{4}\left[\frac{2\cos\left(-\omega t + kD\right)}{D} + \frac{a^2 k r^2 \sin\left(-\omega t + kD\right)}{D^4}\right]$$

Note that the term on the right side dies at least as fast as D^2 due to the r term in the numerator. So at long distances, the left term will dominate, so we can calculate the fields using

$$\vec{A}\left(\vec{x},t\right) = \frac{\mu_0 I_0 a \hat{\theta}}{2D}\cos\left(-\omega t + kD\right).$$ Now this gives:

$$\vec{E} = -\frac{\partial \vec{A}}{\partial t} = -\frac{\mu_0 I_0 a \omega \hat{\theta}}{2D}\sin\left(-\omega t + kD\right)$$

$$\vec{B} = \vec{\nabla}\times\vec{A} = \hat{r}\frac{\mu_0 I_0 a z\left[\cos\left(\omega t - kD\right) - kD\sin\left(\omega t - kD\right)\right]}{2rD^3} + \hat{z}\frac{\mu_0 I_0 a\left[z^2\cos\left(\omega t - kD\right) + kDr^2\sin\left(\omega t - kD\right)\right]}{2rD^3}$$

Taking $kD \gg 1$ gives:

$$\vec{B} = \hat{r}\frac{-ka\mu_0 I_0 z \sin\left(\omega t - kD\right)}{2rD^2} + \hat{z}\frac{ka\mu_0 I_0 r \sin\left(\omega t - kD\right)}{2D^2}.$$

Solution 6.36

Problem 6.36

a. Note first that the vector potential from a single dipole is:

$$\vec{A}\left(\vec{r},t\right) = \frac{\mu_0}{4\pi}\left(-i\omega\vec{p}\right)\frac{e^{ikr}}{r}$$

Let the dipoles have a maximum displacement d and oscillate with a frequency ω. Then the actual moment of the dipoles at time t will be $p\cos(\omega t)\hat{z}$. In this case,

$$r_+ = \left|\vec{r} + \frac{d}{2}\hat{z}\right| \qquad r_- = \left|\vec{r} - \frac{d}{2}\hat{z}\right|$$

$$\vec{A}\left(\vec{r},t\right) = \frac{\mu_0}{4\pi}\left(-i\omega\vec{p}\right)\left[\frac{e^{ikr_-}}{r_-} - \frac{e^{ikr_+}}{r_+}\right]$$

Making the far-zone approximations:

$$r_+ \approx r + \frac{d\cos\theta}{2} + \dots$$

$$\frac{1}{r_+} = \left(r^2 + rd\cos\theta + \frac{d^2}{4}\right)^{-\frac{1}{2}} \approx \frac{1}{2}\left(1 - \frac{d\cos\theta}{2r}\right) + \dots$$

$$r_- \approx r - \frac{d\cos\theta}{2} + \dots$$

$$\frac{1}{r_-} = \left(r^2 - rd\cos\theta + \frac{d^2}{4}\right)^{-\frac{1}{2}} \approx \frac{1}{2}\left(1 + \frac{d\cos\theta}{2r}\right) + \dots$$

Now substituting in the exponential and the denominator, have

$$\vec{A}\left(\vec{r},t\right) \approx \frac{\mu_0}{4\pi}\left(-i\omega\,\vec{p}\,\right)\left[\frac{e^{ikr}}{r}\right](2ikd\cos\theta) = \hat{z}\,\frac{\mu_0}{4\pi}\frac{\omega p_z kd\cos\theta e^{ikr}}{r}.$$

b. $\vec{E} = c\left(\vec{B}\times\vec{n}\right)$ $\vec{n} = \hat{r}$

$$\vec{B} = \vec{\nabla}\times\vec{A} = \frac{id\mu_0}{4\pi r}\cos\theta\sin\theta e^{ikr}\,\hat{\phi}$$

Solution 6.37

Problem 6.37

$$x = l\cos\alpha \quad dx = dl\cos\alpha \quad y = l\sin\alpha \quad dy = dl\sin\alpha$$

$$P_x = \int_0^L \lambda l\cos\alpha\,dl = \frac{\lambda L^2}{2}\cos\alpha \quad P_y = \int_0^L \lambda l\sin\alpha\,dl = \frac{\lambda L^2}{2}\sin\alpha \quad P_z = 0$$

$$Q_{i,j} = \int \rho(\vec{x})\left(3x_i x_j - r^2\delta_{i,j}\right)d^3x$$

$$Q_{x,x} = \int_0^L \lambda\left(3(l\cos\alpha)^2 - l^2\right)dl = L^3\cos^2\alpha - \frac{L^3}{3}$$

$$Q_{y,y} = \int_0^L \lambda\left(3(l\sin\alpha)^2 - l^2\right)dl = L^3\sin^2\alpha - \frac{L^3}{3}$$

$$Q_{z,z} = \int_0^L \lambda\left(3(-l^2)\right)dl = -\frac{L^3}{3}$$

$$Q_{x,y} = Q_{y,x} = \int_0^L \lambda\left(3l^2\sin\alpha\cos\alpha\right)dl = L^3\sin\alpha\cos\alpha$$

$$Q_{x,z} = Q_{z,x} = Q_{y,z} = Q_{z,y} = 0$$

Solution 6.38

Problem 6.38

$$\vec{B} = \vec{\nabla} \times \left[\frac{1}{r^2}\left(\vec{r}\cdot\vec{a}\right)\vec{r} - \frac{1}{r^2}\left(\vec{r}\cdot\vec{r}\right)\vec{a} \right] = \vec{\nabla} \times \left[\frac{1}{r^2}\left(\vec{r}\cdot\vec{a}\right)\vec{r} \right]$$

$$= \varepsilon_{i,j,k}\left(\frac{\partial}{\partial r_j} \frac{1}{\sum_m r_m r_m} \vec{r}_l \vec{a}_l \vec{r}_k \right) = \frac{1}{r^2}\varepsilon_{i,j,k}\left(\vec{a}_j \vec{r}_k\right) = \frac{\vec{a} \times \vec{r}}{r^2}$$

The first term has died off because it has the same arguments, and the last one has died because the antisymmetric operator is zero when j = k.

Solution 6.39

Problem 6.39

a. $U = \int \rho\left(\vec{x}\right)\phi\left(\vec{x}\right)d^3x$

Now bring dipole 2 into the region of dipole 1's field. Place dipole 1 at 0 and define $\vec{R} = \vec{r}_2 - \vec{r}_1$. Now expand this to get

$$U = \int \left[q_2\phi_1(0) + \vec{p}_2\delta^3\left(\vec{R}\right)\vec{\nabla}\phi_1\left(\vec{R}\right) + ... \right]d^3x = -\vec{p}_2\vec{E}_1\left(\vec{R}\right) = \frac{R^2\left(\vec{p}_1\cdot\vec{p}_2\right) - 3\left(\vec{p}_1\cdot\vec{R}\right)\left(\vec{p}_2\cdot\vec{R}\right)}{4\pi\varepsilon_0 R^5}$$

b. $F = -\vec{\nabla}\cdot U = \dfrac{-\left(\vec{p}_1\cdot\vec{p}_2\right)}{4\pi\varepsilon_0}\vec{\nabla}\cdot\dfrac{1}{R^3} + \dfrac{3\vec{\nabla}}{4\pi\varepsilon_0}\cdot\dfrac{\left(\vec{p}_1\cdot\vec{R}\right)\left(\vec{p}_2\cdot\vec{R}\right)}{R^5}$

$$= \frac{\left(\vec{p}_1\cdot\vec{p}_2\right)}{4\pi\varepsilon_0}\frac{1}{R^4}\hat{r} + \frac{3}{4\pi\varepsilon_0}\left[-10x\frac{\left(\vec{p}_1\cdot\vec{R}\right)\left(\vec{p}_2\cdot\vec{R}\right)}{R^6} + \vec{p}_{1x}\frac{\left(\vec{p}_2\cdot\vec{R}\right)}{R^5} + \vec{p}_{2x}\frac{\left(\vec{p}_1\cdot\vec{R}\right)}{R^5} \right]\hat{x} +$$

$$\left[-10y\frac{\left(\vec{p}_1\cdot\vec{R}\right)\left(\vec{p}_2\cdot\vec{R}\right)}{R^6} + \vec{p}_{1y}\frac{\left(\vec{p}_2\cdot\vec{R}\right)}{R^5} + \vec{p}_{2y}\frac{\left(\vec{p}_1\cdot\vec{R}\right)}{R^5} \right]\hat{y}$$

$$+ \left[-10z\frac{\left(\vec{p}_1\cdot\vec{R}\right)\left(\vec{p}_2\cdot\vec{R}\right)}{R^6} + \vec{p}_{1z}\frac{\left(\vec{p}_2\cdot\vec{R}\right)}{R^5} + \vec{p}_{2z}\frac{\left(\vec{p}_1\cdot\vec{R}\right)}{R^5} \right]\hat{z}$$

$$\text{c. } U = \frac{R^2 \left| \vec{p}_1 \right| \left| \vec{p}_2 \right| + 3\frac{\left| \vec{p}_2 \right|}{\left| \vec{p}_1 \right|} \left(\vec{p}_1 \cdot \vec{R} \right)^2}{4\pi\varepsilon_0 R^5}$$

Need to minimize the dot product, which means that the dipole's direction should be perpendicular to the other dipole.

Solution 6.40

Problem 6.40

a. First, find an expression for the overall vector potential; it is critical to have Q distributed evenly over the sphere. Let $\sigma(r) = \frac{3Q}{4\pi R^3}$. Now integrate to find the vector potential at a particular r inside the sphere.

$$\vec{A} = \int_0^r A_{out} + \int_r^R A_{in} = \int_0^r \frac{\mu_0 x^4 \omega\sigma}{3} \left(\frac{\sin\theta}{r^2} \right) \widehat{\phi}\, dx + \int_r^R \frac{\mu_0 x \omega\sigma}{3} (r\sin\theta) \widehat{\phi}\, dx = \frac{\mu_0 \omega\sigma}{30} \left[5R^2 - 3r^2 \right] (r\sin\theta)\, \widehat{\phi}$$

Now the expression for B becomes relatively simple, because A is only in the phi direction:

$$\vec{B} = \vec{\nabla} \times \vec{A} = \hat{r}\frac{\mu_0 \omega\sigma\cos\theta}{15}\left[5R^2 - 3r^2 \right] - \hat{\theta}\frac{\mu_0 \omega\sigma\sin\theta}{15}\left[5R^2 - 6r^2 \right] = \hat{z}\frac{\mu_0 \omega\sigma\cos\theta}{15}\left[5R^2 - 3r^2 \right] + \hat{\theta}\frac{\mu_0 \omega\sigma\sin\theta}{15}\left[3r^2 \right]$$

b. $\vec{m} = \frac{1}{2}\int \left(\vec{x} \times \vec{J}(\vec{x}) \right) d^3x = \frac{\hat{z}}{2}\int r\vec{J}(\vec{x})\sin\theta d^3x = \frac{\hat{z}}{2}\int r^2 \sigma\omega\sin^2\theta d^3x$

$$= \frac{\hat{z}}{2}\iiint r^2 \sigma\omega\sin^2\theta r^2\sin\theta dr d\theta d\phi = \hat{z}\frac{4}{15}R^5\pi\sigma\omega$$

c. $\vec{B}_{avg} = \frac{3}{4\pi R^3}\iiint \left\{ \hat{z}\frac{\mu_0 \omega\sigma\cos\theta}{15}\left[5R^2 - 3r^2 \right] + \hat{\theta}\frac{\mu_0 \omega\sigma\sin\theta}{15}\left[3r^2 \right] \right\} r^2\sin\theta dr d\theta d\phi$

But note that because of azimuthal symmetry, the portion in the azimuthal direction will contribute only to z, as the sine of the angle:

$$\vec{B}_{avg} = \frac{3}{4\pi R^3}\iiint \left\{ \hat{z}\frac{\mu_0 \omega\sigma\cos\theta}{15}\left[5R^2 - 3r^2 \right] - \hat{z}\sin\theta\frac{\mu_0 \omega\sigma\sin\theta}{15}\left[3r^2 \right] \right\} r^2\sin\theta dr d\theta d\phi = \frac{2R^2 \mu_0 \omega\sigma}{15}\hat{z} = \frac{\mu_0}{2\pi R^2}\vec{m}$$

Solution 6.41

a. The form of the fields for a boost in particular dimension is

$$\frac{E'^1}{c} = \frac{E^1}{c} \quad B'^1 = B^1$$

$$\frac{E'^2}{c} = \gamma\left(\frac{E^2}{c} - \beta B^3\right) \quad B'^2 = \gamma\left(\beta\frac{E^3}{c} + B^2\right)$$

$$\frac{E'^3}{c} = \gamma\left(\frac{E^3}{c} + \beta B^3\right) \quad B'^3 = \gamma\left(-\beta\frac{E^2}{c} + B^3\right)$$

So, to eliminate the electric field in this way, it is necessary to take a boost in the direction perpendicular to the field. Recall that $\beta = \frac{c}{v}$, so since $\left|\vec{E}\right| = c\left|\vec{B}\right|$, the boost would have to be at speed c: this would affect γ so that both fields would be brought to zero in all directions except the boosted one. For boosts not perpendicular to the magnetic field, the boost speed would have to be faster than light. Thus, it is not possible.

b. Let $E \propto \hat{x}_2, B \propto \hat{x}_3$. Then

$$\frac{E'^1}{c} = B'^1 = 0$$

$$\frac{E'^2}{c} = \frac{1}{\sqrt{1-\frac{v^2}{c^2}}}\left(\frac{E^2}{c} - \frac{v}{c}B^3\right) = \frac{1}{\sqrt{1-\frac{v^2}{c^2}}}\left(B^2 - \frac{v}{c}B^3\right)$$

$$B'^3 = \frac{1}{\sqrt{1-\frac{v^2}{c^2}}}\left(-\frac{v}{c}B^2 + B^3\right)$$

Perpendicular: yes. $\left|\vec{E}\right| = c\left|\vec{B}\right|$: yes, because $\frac{\partial}{\partial t}\left(\frac{\vec{E}^2}{c^2} - \vec{B}^2\right) = 0$.

c. $\frac{\partial p_x}{\partial t} = q\left(v_y B\right) \quad \frac{\partial p_y}{\partial t} = q(qE - v_x B) \quad \frac{\partial p_z}{\partial t} = 0 \quad \frac{\partial E}{\partial t} = q\left(v_y cB\right)$

d. In order to be conserved, they must be independent of time.

$$\frac{\partial p^\mu}{\partial \tau} = qF^{\mu v}u_v = qF^{\mu v}g_{v\alpha}u^\alpha$$

$$F = \begin{pmatrix} 0 & 0 & -B & 0 \\ 0 & 0 & -B & 0 \\ B & B & 0 & 0 \\ 0 & 0 & 0 & 0 \end{pmatrix}$$

$$\frac{\partial p_z}{\partial t} = 0$$

$$p_x - \frac{mc}{\sqrt{1 - \frac{v^2}{c^2}}} = \frac{\partial p_x}{\partial \tau} - \frac{\partial p_0}{\partial \tau} = -\frac{qB}{m}v_y + \frac{qB}{m}v_y = 0$$

Solution 6.42

Problem 6.42

a. Consider an infinitesimal change in position within the particle's rest frame:

$$\frac{\partial E}{\partial t} = q\vec{v} \cdot \vec{E}$$

$$\frac{\partial}{\partial t}(mc^2\gamma) = q\vec{v} \cdot \vec{E} = q\vec{v} \cdot (-\nabla\phi) = -q\frac{d\phi}{dt}$$

$$\Rightarrow \frac{d}{dt}(mc^2\gamma + q\phi) = 0.$$

b. Under the spherically symmetric potential, angular momentum is conserved. Further, total energy is conserved as shown above.

$$E = mc^2\gamma + q\phi \quad J = L = \vec{r} \times \vec{p} = m\gamma r^2\dot{\theta} \quad \frac{dE}{dt} = \frac{dL}{dt} = 0$$

$$\vec{p}^2 = m^2\gamma^2\vec{v}^2 = m^2\gamma^2(\dot{r}^2 + r^2\dot{\theta}^2) = m^2\gamma^2\dot{r}^2 + \frac{J^2}{r^2}$$

$$E = c\sqrt{\vec{p}^2 + m^2c^2} + \frac{qe'}{r} \Rightarrow \left(E - \frac{qe'}{r}\right)^2 = \vec{p}^2c^2 + m^2c^4 = m^2c^2\gamma^2\dot{r}^2 + \frac{J^2c^2}{r^2} + m^2c^4$$

$$u = \frac{1}{r} \quad \dot{r} = -\frac{1}{u^2}\frac{du}{d\theta}\frac{d\theta}{dt} = -\frac{1}{u^2}u'\dot{\theta} = \frac{-Ju'}{m\gamma}$$

Combining these results gives:

$c^2J^2\left(u'^2 + u^2\right) - 2Eqeu' + q^2e'^2u^2 = E^2 - m^2c^4$, which may be written

$$a\left(u'^2 + u^2\right) - bu + cu^2 = d$$

Differentiating gives the equation of motion:

$$a(2u'u'' + 2uu') - bu' + 2cuu' = 0$$

$$a(2u'' + 2u) - b + 2cu = 0$$

which gives

$$u(\theta) = \frac{b}{2(a+c)} + C_1 \exp\left(\sqrt{\frac{-a-c}{a}}\theta\right) + C_2 \exp\left(-\sqrt{\frac{-a-c}{a}}\theta\right).$$

This is not necessarily periodic, so the relativistic solution can precess about the center. Depending upon a and c, the solution may or may not be bound.

References
Electricity and magnetism

1. Burns, J. A., Lamy, P. L.& Soter, S., "Radiation forces on small particles in the solar-system," *Icarus* **40**, 1-48 (1979).

2. Jackson, J. D., *Classical Electrodynamics*, Wiley, New York, NY (1998).

3. Jacobus, R. W., "Uncertainties in satellite position due to solar radiation pressure effects," *Technical Documentary Report* **ESD-TDR-64-147**, (1964).

4. Larmor, J., "On a dynamical theory of the electric and luminiferous medium," *Philosophical Transactions of the Royal Society* **190**, 205-300 (1897).

5. Robertson, H. P., "Dynamical effects of radiation in the solar system," *Monthly Notices of the Royal Astronomical Society* **97**, 0423-0438 (1936).

Made in the USA
Monee, IL
08 January 2025